Lecture Notes in Artificia

Edited by R. Goebel, J. Siekmann, a

Subseries of Lecture Notes in Computer Science

FoLLI Publications on Logic, Language and Information

Philippe de Groote   Markus Egg
Laura Kallmeyer (Eds.)

Formal Grammar

14th International Conference, FG 2009
Bordeaux, France, July 25-26, 2009
Revised Selected Papers

 Springer

Series Editors

Randy Goebel, University of Alberta, Edmonton, Canada
Jörg Siekmann, University of Saarland, Saarbrücken, Germany
Wolfgang Wahlster, DFKI and University of Saarland, Saarbrücken, Germany

Volume Editors

Philippe de Groote
INRIA Nancy - Grand Est
615 Rue du Jardin Botanique, 54600 Villers-lès-Nancy, France
E-mail: philippe.de.groote@loria.fr

Markus Egg
Humboldt-Universität zu Berlin, Institut für Anglistik und Amerikanistik
Unter den Linden 6, 10117 Berlin, Germany
E-mail: markus.egg@rz.hu-berlin.de

Laura Kallmeyer
Heinrich-Heine-Universität Düsseldorf, Institut für Sprache und Information
Universitätsstraße 1, 40225 Düsseldorf, Germany
E-mail: kallmeyer@phil.uni-duesseldorf.de

ISSN 0302-9743   e-ISSN 1611-3349
ISBN 978-3-642-20168-4   e-ISBN 978-3-642-20169-1
DOI 10.1007/978-3-642-20169-1
Springer Heidelberg Dordrecht London New York

Library of Congress Control Number: 2011923662

CR Subject Classification (1998): F.4.2, F.4, F.1, I.1, I.2.7

LNCS Sublibrary: SL 7 – Artificial Intelligence

Typesetting: Camera-ready by author, data conversion by Scientific Publishing Services, Chennai, India

Printed on acid-free paper

Springer is part of Springer Science+Business Media (www.springer.com)

Preface

The Formal Grammar conference series provides a forum for the presentation of new and original research on formal grammar, mathematical linguistics and the application of formal and mathematical methods to the study of natural language.

FG-2009, the 14th conference on Formal Grammar, was held in Bordeaux, France during July 25–26, 2009. The conference consisted of 13 contributed papers (selected out of 26 submissions), and two invited talks.

We would like to thank the people who made this 14th FG conference possible: the two invited speakers, Makoto Kanazawa and Gregory M. Kobele, the members of the Program Committee, and the members of the Organizing Committee.

April 2010

Philippe de Groote
Markus Egg
Laura Kallmeyer

Organization

Program Committee

Wojciech Buszkowski	Adam Mickiewicz University, Poland
Berthold Crysmann	University of Bonn, Germany
Alexandre Dikovsky	University of Nantes, France
Denys Duchier	University of Orléans, France
Markus Egg (Co-chair)	Humboldt-Universität Berlin, Germany
Annie Foret	IRISA - IFSIC, France
Nissim Francez	Technion, Israel
Gerhard Jaeger	University of Tübingen, Germany
Laura Kallmeyer (Co-chair)	University of Düsseldorf, Germany
Makoto Kanazawa	National Institute of Informatics, Japan
Stephan Kepser	University of Tübingen, Germany
Marco Kuhlmann	Uppsala University, Sweden
Glyn Morrill	UPC. Barcelona Tech, Spain
Richard Moot	University of Bordeaux 3, France
Larry Moss	Indiana University, USA
Stefan Müller	Free University of Berlin, Germany
Mark-Jan Nederhof	University of St. Andrews, UK
Joakim Nivre	Växjö University, Sweden
Frank Richter	University of Tübingen, Germany
Ed Stabler	UCLA, USA
Hans-Jörg Tiede	Illinois Wesleyan University, USA
Jesse Tseng	CNRS - CLLE-ERSS, France
Shuly Wintner	University of Haifa, Israel

Organizing Committee

Nicolas Alcaraz	INRIA Nancy - Grand Est, France
Anne-Lise Charbonnier	INRIA Nancy - Grand Est, France
Philippe de Groote (Chair)	INRIA Nancy - Grand Est, France
Laetitia Grimaldi	INRIA Bordeaux - Sud Ouest, France

FG Standing Committee

Philippe de Groote	INRIA Nancy - Grand Est, France
Markus Egg	Humboldt-Universität Berlin, Germany
Laura Kallmeyer	University of Tübingen, Germany
Gerald Penn	University of Toronto, Canada

Table of Contents

Contributed Papers

Linear Conjunctive Grammars and One-Turn Synchronized Alternating Pushdown Automata

Tamar Aizikowitz and Michael Kaminski

Department of Computer Science
Technion – Israel Institute of Technology
Haifa 32000, Israel

Abstract. In this paper we introduce a sub-family of synchronized alternating pushdown automata, *one-turn Synchronized Alternating Pushdown Automata*, which accept the same class of languages as those generated by Linear Conjunctive Grammars. This equivalence is analogous to the classical equivalence between one-turn PDA and Linear Grammars, thus strengthening the claim of Synchronized Alternating Pushdown Automata as a natural counterpart for Conjunctive Grammars.

1 Introduction

Context-free languages lay at the very foundation of Computer Science, proving to be one of the most appealing language classes for practical applications. One the one hand, they are quite expressive, covering such syntactic constructs as necessary, e.g., for mathematical expressions. On the other hand, such languages are polynomially parsable, making them practical for real world applications. However, research in certain fields has raised a need for computational models which extend context-free models, without losing their computational efficiency.

Conjunctive Grammars (CG) are an example of such a model. Introduced by Okhotin in [10], CG are a generalization of context-free grammars which allow explicit intersection operations in rules, thereby adding the power of conjunction. CG were shown by Okhotin to accept all finite conjunctions of context-free languages, as well as some additional languages. However, there is no known nontrivial technique to prove a language cannot be derived by a CG, so their exact placing in the Chomsky hierarchy is unknown. Okhotin proved the languages generated by these grammars to be polynomially parsable [10,11], making the model practical from a computational standpoint, and therefore, of interest for applications in various fields such as, programming languages.

Alternating automata models were first introduced by Chandra, Kozen and Stockmeyer in [3]. Alternating Pushdown Automata (APDA) were further explored in [9], and shown to accept exactly the exponential time languages. As such, they are too strong a model for Conjunctive Grammars. Synchronized Alternating Pushdown Automata (SAPDA), introduced in [1], are a weakened version of Alternating Pushdown Automata, which accept conjunctions of context-free languages. In [1], SAPDA were proven to be equivalent to CG[1].

---

[1] We call two models equivalent if they accept/generate the same class of languages.

P. de Groote, M. Egg, and L. Kallmeyer (Eds.): FG 2009, LNAI 5591, pp. 1–16, 2011.

In [10], Okhotin defined a sub-family of Conjunctive Grammars called *Linear Conjunctive Grammars* (LCG), analogously to the definition of Linear Grammars as a sub-family of Context-free Grammars. LCG are an interesting sub-family of CG as they have especially efficient parsing algorithms, see [12], making them particularly appealing from a computational standpoint. Also, many of the interesting languages derived by Conjunctive Grammars, can in fact be derived by Linear Conjunctive Grammars. In [13], Okhotin proved that LCG are equivalent to a type of *Trellis Automata*.

It is well-known, due to Ginsburg and Spanier [5], that Linear Grammars are equivalent to one-turn PDA. One-turn PDA are a sub-family of pushdown automata, where in each computation the stack height switches only once from non-decreasing to non-increasing. That is, once a transition replaces the top symbol of the stack with ϵ, all subsequent transitions may write at most one character.

In this paper we introduce a sub-family of SAPDA, *one-turn Synchronized Alternating Pushdown Automata*, and prove that they are equivalent to Linear Conjunctive Grammars. The equivalence is analogous to the classical equivalence between one-turn PDA and Linear Grammars. This result greatly strengthens the claim of SAPDA as a natural automaton counterpart for Conjunctive Grammars.

The paper is organized as follows. In Section 2 we recall the definitions of Conjunctive Grammars, Linear Conjunctive Grammars, and SAPDA. In Section 3 we introduce one-turn SAPDA as a sub-family of general SAPDA. Section 4 details our main result, namely the equivalence of the LCG and one-turn SAPDA models. Section 5 discusses the relationship between LCG and Mildly Context Sensitive Languages, and Section 6 is a short conclusion of our work.

2   Preliminaries

In the following section, we repeat the definitions of Conjunctive Grammars, Linear Conjunctive Grammars, and Synchronized Alternating Pushdown Automata, as presented in [1].

2.1   Conjunctive Grammars

The following definitions are taken from [10].

Definition 1. *A* Conjunctive Grammar *is a quadruple* $G = (V, \Sigma, P, S)$, *where*

- V, Σ *are disjoint finite sets of non-terminal and terminal symbols respectively.*
- $S \in V$ *is the designated start symbol.*
- *P is a finite set of rules of the form* $A \to (\alpha_1 \,\&\, \cdots \,\&\, \alpha_k)$ *such that* $A \in V$ *and* $\alpha_i \in (V \cup \Sigma)^*$, $i = 1, \ldots, k$. *If* $k = 1$ *then we write* $A \to \alpha$.

Definition 2. Conjunctive Formulas *over* $V \cup \Sigma \cup \{(,), \&\}$ *are defined by the following recursion.*

- ϵ is a conjunctive formula.
- Every symbol in $V \cup \Sigma$ is a conjunctive formula.
- If $\mathcal{A}$ and $\mathcal{B}$ are formulas, then $\mathcal{A}\mathcal{B}$ is a conjunctive formula.
- If $\mathcal{A}_1, \ldots, \mathcal{A}_k$ are formulas, then $(\mathcal{A}_1 \& \cdots \& \mathcal{A}_k)$ is a conjunctive formula.

Notation. Below we use the following notation: σ, τ, etc. denote terminal symbols, u, w, y, etc. denote terminal words, A, B, etc. denote non-terminal symbols, α, β, etc. denote non-terminal words, and $\mathcal{A}$, $\mathcal{B}$, etc. denote conjunctive formulas. All the symbols above may also be indexed.

Definition 3. *For a conjunctive formula* $\mathcal{A} = (\mathcal{A}_1 \& \cdots \& \mathcal{A}_k)$, $k \geq 2$, *the* $\mathcal{A}_i$s, $i = 1, \ldots, k$, *are called* conjuncts *of* $\mathcal{A}$,[2] *and* $\mathcal{A}$ *is called the* enclosing formula. *If* $\mathcal{A}_i$ *contains no &s, then it is called a* simple conjunct.[3]

Definition 4. *For a CG G, the relation of* immediate derivability *on the set of conjunctive formulas,* $\Rightarrow_G$, *is defined as follows.*

(1) $s_1 A s_2 \Rightarrow_G s_1(\alpha_1 \& \cdots \& \alpha_k)s_2$ *for all* $A \to (\alpha_1 \& \cdots \& \alpha_k) \in P$, *and*
(2) $s_1(w\& \cdots \&w)s_2 \Rightarrow_G s_1 w s_2$ *for all* $w \in \Sigma^*$,

where $s_1, s_2 \in (V \cup \Sigma \cup \{(,), \&\})^*$. *As usual,* $\Rightarrow_G^*$ *is the reflexive and transitive closure of* $\Rightarrow_G$,[4] *the language* $L(\mathcal{A})$ *of a conjunctive formula* $\mathcal{A}$ *is*

$$L(\mathcal{A}) = \{w \in \Sigma^* : \mathcal{A} \Rightarrow_G^* w\},$$

and the language $L(G)$ *of G is*

$$L(G) = \{w \in \Sigma^* : S \Rightarrow_G^* w\}.$$

We refer to (1) as production *and (2) as* contraction *rules, respectively.*

Example 1. ([10, Example 1]) The following conjunctive grammar generates the non-context-free language $\{a^n b^n c^n : n = 0, 1, \ldots\}$, called the *multiple agreement* language. $G = (V, \Sigma, P, S)$, where

- $V = \{S, A, B, C, D, E\}$, $\Sigma = \{a, b, c\}$, and
- P consists of the following rules:
    $S \to (C \& A)$; $C \to Cc \mid D$; $A \to aA \mid E$; $D \to aDb \mid \epsilon$; $E \to bEc \mid \epsilon$

The intuition of the above derivation rules is as follows.

$$L(C) = \{a^m b^m c^n : m, n = 0, 1, \ldots\},$$

$$L(A) = \{a^m b^n c^n : m, n = 0, 1, \ldots\},$$

---

[2] Note that this definition is different from Okhotin's definition in [10].
[3] Note that, according to this definition, conjunctive formulas which are elements of $(V \cup \Sigma)^*$ are not simple conjuncts or enclosing formulas.
[4] In particular, a terminal word w is derived from a conjunctive formula $(\mathcal{A}_1 \& \cdots \& \mathcal{A}_k)$ if and only if it is derived from each $\mathcal{A}_i$, $i = 1, \ldots, k$.

and

$$L(G) = L(C) \cap L(A) = \{a^n b^n c^n : n = 0, 1, \ldots\}.$$

For example, the word $aabbcc$ can be derived as follows.

$$S \Rightarrow (C \ \& \ A) \Rightarrow (Cc \ \& \ A) \Rightarrow (Ccc \ \& \ A) \Rightarrow (Dcc \ \& \ A) \Rightarrow (aDbcc \ \& \ A)$$
$$\Rightarrow (aaDbbcc \ \& \ A) \Rightarrow (aabbcc \ \& \ A) \Rightarrow (aabbcc \ \& \ aA)$$
$$\Rightarrow (aabbcc \ \& \ aaA) \Rightarrow (aabbcc \ \& \ aaE) \Rightarrow (aabbcc \ \& \ aabEc)$$
$$\Rightarrow (aabbcc \ \& \ aabbEcc) \Rightarrow (aabbcc \ \& \ aabbcc) \Rightarrow aabbcc$$

2.2   Linear Conjunctive Grammars

Okhotin defined in [10] a sub-family of conjunctive grammars called *Linear Conjunctive Grammars* (LCG) and proved in [13] that they are equivalent to *Trellis Automata*.[5] The definition of LCGs is analogous to the definition of Linear Grammars as a sub-family of Context-free Grammars.

Definition 5. *A conjunctive grammar* $G = (V, \Sigma, P, S)$ *is said to be* linear *if all rules in* P *are in one of the following forms.*

- $A \rightarrow (u_1 B_1 v_1 \ \& \ \cdots \ \& \ u_k B_k v_k); u_i, v_i \in \Sigma^*$ *and* $A, B_i \in V$, *or*
- $A \rightarrow w; w \in \Sigma^*$, *and* $A \in V$.

Several interesting languages can be generated by LCGs. In particular, the grammar in Example 1 is linear. The following is a particularly interesting example, due to Okhotin, of a Linear CG which uses recursive conjunctions to derive a language which *cannot* be obtained by a finite conjunction of context-free languages.

Example 2. ([10, Example 2]) The following linear conjunctive grammar derives the non-context-free language $\{w\$w : w \in \{a, b\}^*\}$, called *reduplication* with a center marker. $G = (V, \Sigma, P, S)$, where

- $V = \{S, A, B, C, D, E\}$, $\Sigma = \{a, b, \$\}$, and
- P consists of the following derivation rules.

$$
\begin{array}{ll}
S \rightarrow (C \ \& \ D) \quad ; & C \rightarrow aCa \mid aCb \mid bCa \mid bCb \mid \$ \\
D \rightarrow (aA \ \& \ aD) \mid (bB \ \& \ bD) \mid \$E \quad ; & A \rightarrow aAa \mid aAb \mid bAa \mid bAb \mid \$Ea \\
B \rightarrow aBa \mid aBb \mid bBa \mid bBb \mid \$Eb \quad ; & E \rightarrow aE \mid bE \mid \epsilon
\end{array}
$$

The non-terminal C verifies that the lengths of the words before and after the center marker $\$$ are equal. The non-terminal D derives the language $\{w\$uw|w, u \in \{a, b\}\}$. The grammar language is the intersection of these two languages, i.e., the reduplication with a center marker language. For a more detailed description, see [10, Example 2].

---

[5] As Trellis Automata are not a part of this paper, we omit the definition, which can be found in [4] or [13].

2.3   Synchronized Alternating Pushdown Automata

Next, we recall the definition of *Synchronized Alternating Pushdown Automata* (SAPDA). Introduced in [1], SAPDA are a variation on standard PDA which add the power of conjunction. In the SAPDA model, transitions are made to a conjunction of states. The model is non-deterministic, therefore, several different conjunctions of states may be possible from a given configuration. If all conjunctions are of one state only, the automaton is a standard PDA.[6]

The stack memory of an SAPDA is a tree. Each leaf has a processing head which reads the input and writes to its branch independently. When a multiple-state conjunctive transition is applied, the stack branch splits into multiple branches, one for each conjunct.[7] The branches process the input independently, however sibling branches must empty synchronously, after which the computation continues from the parent branch.

Definition 6. *A* Synchronized Alternating Pushdown Automaton *is a tuple* $A = (Q, \Sigma, \Gamma, \delta, q_0, \bot)$, *where* δ *is a function that assigns to each element of* $Q \times (\Sigma \cup \{\epsilon\}) \times \Gamma$ *a finite subset of*

$$\{(q_1, \alpha_1) \wedge \cdots \wedge (q_k, \alpha_k) : k = 1, 2, \ldots, \ i = 1, \ldots, k, \ q_i \in Q, \ and \ \alpha_i \in \Gamma^*\}.$$

Everything else is defined as in the standard PDA model. Namely,

- Q *is a finite set of states,*
- Σ *and* Γ *are the input and the stack alphabets, respectively,*
- $q_0 \in Q$ *is the initial state, and*
- $\bot \in \Gamma$ *is the initial stack symbol,*

see, e.g., [6, pp. 107–112].

We describe the current stage of the automaton computation as a labeled tree. The tree encodes the stack contents, the current states of the stack-branches, and the remaining input to be read for each stack-branch. States and remaining inputs are saved in leaves only, as these encode the stack-branches currently being processed.

Definition 7. *A* configuration *of an SAPDA is a labeled tree. Each internal node is labeled* $\alpha \in \Gamma^*$ *denoting the stack-branch contents, and each leaf node is labeled* (q, w, α), *where*

- $q \in Q$ *is the current state,*
- $w \in \Sigma^*$ *is the remaining input to be read, and*
- $\alpha \in \Gamma^*$ *is the stack-branch contents.*

---

[6] This type of formulation for alternating automata models is equivalent to the one presented in [3], and is standard in the field of Formal Verification, e.g., see [7].

[7] This is similar to the concept of a transition from a universal state in the standard formulation of alternating automata, as all branches must accept.

For a node v in a configuration T, we denote the label of v in T by $T(v)$. If a configuration has a single node only,[8] it is denoted by the label of that node. That is, if a configuration T has a single node labeled (q, w, α), then T is denoted by (q, w, α).

At each computation step, a transition is applied to one stack-branch.[9] If a branch empties, it cannot be chosen for the next transition (because it has no top symbol). If all sibling branches are empty, and each branch emptied with the *same* remaining input (i.e., after processing the same portion of the input) and with the same state, the branches are collapsed back to the parent branch.

Definition 8. *Let A be an SAPDA and let T, T' be configurations of A. We say that T yields T' in one step, denoted $T \vdash_A T'$ (A is omitted if understood from the context), if one of the following holds.*

- *There exists a leaf node v in T, $T(v) = (q, \sigma w, X\alpha)$ and a transition $(q_1, \alpha_1) \wedge \cdots \wedge (q_k, \alpha_k) \in \delta(q, \sigma, X)$ which satisfy the conditions below.*
  - *If $k = 1$, then T' is obtained from T by relabeling v with $(q_1, w, \alpha_1\alpha)$.*
  - *If $k > 1$, then T' is obtained from T by relabeling v with α, and adding to it k child nodes $v_1, \ldots, v_k$ such that $T'(v_i) = (q_i, w, \alpha_i)$, $i = 1, \ldots, k$.*
  In this case we say that the computation step is based on $(q_1, \alpha_1) \wedge \cdots \wedge (q_k, \alpha_k)$ applied to v.
- *There is a node v in T, $T(v) = \alpha$, that has k children $v_1, \ldots, v_k$, all of which are leaves labeled the same (p, w, ϵ), and T' is obtained from T by removing all leaf nodes v_i, $i = 1, \ldots, k$ and relabeling v with (p, w, α).*

In this case we say that the computation step is based on a collapsing of the child nodes of v.

As usual, we denote by $\vdash_A^*$ the reflexive and transitive closure of $\vdash_A$.

Definition 9. *Let A be an SAPDA and let $w \in \Sigma^*$.*

- *The initial configuration of A on w is the configuration $(q_0, w, \bot)$.[10]*
- *An accepting configuration of A is a configuration of the form (q, ϵ, ϵ).*
- *A computation of A on w is a sequence of configurations $T_0, \ldots, T_n$, where*
  - *T_0 is the initial configuration,*
  - *$T_{i-1} \vdash_A T_i$ for $i = 1, \ldots, n$, and*
  - *all leaves v of T_n are labeled (q, ϵ, α), in particular, the entire input string has been read.*
- *An accepting computation of A on w is a computation whose last configuration T_n is accepting.*

---

[8] That is, the configuration tree consists of the root only, which is also the only leaf of the tree.

[9] Equivalently, all branches can take one step together. This formulation simplifies the configurations, as all branches read the input at the same pace. However, it is less correlated with the grammar model, making equivalence proofs more involved.

[10] That is, the initial configuration of A on w is a one node tree whose only node is labeled $(q_0, w, \bot)$.

The language $L(A)$ of A is the set of all $w \in \Sigma^*$ such that A has an accepting computation on w.[11]

Example 3. The SAPDA $A = (Q, \Sigma, \Gamma, \delta, q_0, \bot)$ defined below accepts the non-context-free language

$$\{w : |w|_a = |w|_b = |w|_c\}.$$

- $Q = \{q_0, q_1, q_2\}$,
- $\Sigma = \{a, b, c\}$,
- $\Gamma = \{\bot, a, b, c\}$, and
- δ is defined as follows.
  - $\delta(q_0, \epsilon, \bot) = \{(q_1, \bot) \wedge (q_2, \bot)\}$,
  - $\delta(q_1, \sigma, \bot) = \{(q_1, \sigma\bot)\}$, $\sigma \in \{a, b\}$,
  - $\delta(q_2, \sigma, \bot) = \{(q_2, \sigma\bot)\}$, $\sigma \in \{b, c\}$,
  - $\delta(q_1, \sigma, \sigma) = \{(q_1, \sigma\sigma)\}$, $\sigma \in \{a, b\}$,
  - $\delta(q_2, \sigma, \sigma) = \{(q_2, \sigma\sigma)\}$, $\sigma \in \{b, c\}$,
  - $\delta(q_1, \sigma', \sigma'') = \{(q_1, \epsilon)\}$, $(\sigma', \sigma'') \in \{(a, b), (b, a)\}$,
  - $\delta(q_2, \sigma', \sigma'') = \{(q_2, \epsilon)\}$, $(\sigma', \sigma'') \in \{(b, c), (c, b)\}$,
  - $\delta(q_1, c, X) = \{(q_1, X)\}$, $X \in \{\bot, a, b\}$,
  - $\delta(q_2, a, X) = \{(q_2, X)\}$, $X \in \{\bot, b, c\}$, and
  - $\delta(q_i, \epsilon, \bot) = \{(q_0, \epsilon)\}$, $i = 1, 2$.

The first step of the computation opens two branches, one for verifying that the number of *a*s in the input word equals to the number of *b*s, and the other for verifying that the number of *b*s equals to the number of *c*s. If both branches manage to empty their stack then the word is accepted.

Figure 1 shows the contents of the stack tree at an intermediate stage of a computation on the word *abbcccaab*. The left branch has read *abbccc* and indicates that one more *b*s than *a*s have been read, while the right branch has read *abb* and indicates that two more *b*s than *c*s have been read. Figure 2 shows the configuration corresponding the above computation stage of the automaton.

We now consider the following example of an SAPDA which accepts the non-context-free language $\{w\$uw : w, u \in \{a, b\}\}$. Note that the intersection of this language with $\{u\$v : u, v \in \{a, b\} \wedge |u| = |v|\}$ is the *reduplication with a center marker* language. As the latter language is context-free, and SAPDA are closed under intersection, the construction can easily be modified to accept the reduplication language.

The example is of particular interest as it showcases the model's ability to utilize recursive conjunctive transitions, allowing it to accept languages which are not finite intersections of context-free languages. Moreover, the example gives additional intuition towards understanding Okhotin's grammar for the reduplication language as presented in Example 2. The following automaton accepts the language derived by the non-terminal D in the grammar.

---

[11] Alternatively, one can extend the definition of A with a set of *accepting* states $F \subseteq Q$ and define acceptance by accepting states, similarly to the classical definition. It can readily be seen that such an extension results in an equivalent model of computation.

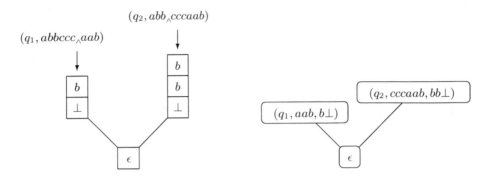

Fig. 1. Intermediate stage of a computation on $abbcccaab$

Fig. 2. The configuration corresponding to Figure 1

Example 4. (cf. Example 2) The SAPDA $A = (Q, \Sigma, \Gamma, \delta, q_0, \bot)$ defined below accepts the non-context-free language

$$\{w\$uw : w, u \in \{a, b\}^*\}.$$

- $Q = \{q_0, q_e\} \cup \{q_\sigma^1 : \sigma \in \{a, b\}\} \cup \{q_\sigma^2 : \sigma \in \{a, b\}\}$,
- $\Sigma = \{a, b, \$\}$,
- $\Gamma = \{\bot, \#\}$, and
- δ is defined as follows.
  1. $\delta(q_0, \sigma, \bot) = \{(q_\sigma^1, \bot) \wedge (q_0, \bot)\}$, $\sigma \in \{a, b\}$
  2. $\delta(q_\sigma^1, \tau, X) = \{(q_\sigma^1, \#X)\}$, $\sigma, \tau \in \{a, b\}$, $X \in \Gamma$,
  3. $\delta(q_0, \$, \bot) = \{(q_e, \epsilon)\}$,
  4. $\delta(q_\sigma^1, \$, X) = \{(q_\sigma^2, X)\}$, $\sigma \in \{a, b\}$, $X \in \Gamma$,
  5. $\delta(q_\sigma^2, \tau, X) = \{(q_\sigma^2, X)\}$, $\sigma, \tau \in \{a, b\} : \sigma \neq \tau$, $X \in \Gamma$,
  6. $\delta(q_\sigma^2, \sigma, X) = \{(q_\sigma^2, X), (q_e, X)\}$, $\sigma \in \{a, b\}$, $X \in \Gamma$,
  7. $\delta(q_e, \sigma, \#) = \{(q_e, \epsilon)\}$, $\sigma \in \{a, b\}$,
  8. $\delta(q_e, \epsilon, \bot) = \{(q_e, \epsilon)\}$

The computations of the automaton have two main phases: before and after the $\$$ sign is encountered in the input. In the first phase, each input letter σ which is read leads to a conjunctive transition (transition 1) that opens two new stack-branches. One new branch continues the recursion, while the second checks that the following condition is met.

Assume σ is the n-th letter from the $\$$ sign. If so, the new stack branch opened during the transition on σ will verify that the n-th letter from the end of the input is also σ. This way, if the computation is accepting, the word will in fact be of the form $w\$uw$. To be able to check this property, the branch must know σ and σ's relative position (n) to the $\$$ sign. To "remember" σ, the state of the branch head is q_σ^1 (the 1 superscript denoting that the computation is in the first phase). To find n, the branch adds a $\#$ sign to its stack for each input character read (transition 2), until the $\$$ is encountered in the input. Therefore, when the

$ is read, the number of #s in the stack branch will be the number of letters between σ and the $ sign in the first half of the input word.

Once the $ is read, the branch perpetuating the recursion ceases to open new branches, and instead transitions to q_e and empties its stack (transition 3). All the other branches denote that they have moved to the second phase of the computation by transitioning to states q_σ^2 (transition 4). From this point onward, each branch "waits" to see the σ encoded in its state in the input (transition 5). Once it does encounter σ, it can either ignore it and continue to look for another σ in the input (in case there are repetitions in w of the same letter), or it can "guess" that this is the σ which is n letters from the end of the input, and move to state q_e (transition 6).

After transitioning to q_e, one # is emptied from the stack for every input character read. If in fact σ was the right number of letters from the end, the $\perp$ sign of the stack branch will be exposed exactly when the last input letter is read. At this point, an ϵ-transition is applied which empties the stack branch (transition 8).

If all branches successfully "guess" their respective σ symbols then the computation will reach a configuration where all leaf nodes are labeled $(q_e, \epsilon, \epsilon)$. From here, successive branch collapsing steps can be applied until an accepting configuration is reached.

Consider a computation on the word $abb\$babb$. Figure 3 shows the contents of the stack tree after all branches have read the prefix ab. The rightmost branch is the branch perpetuating the recursion. The leftmost branch remembers seeing a in the input, and has since counted one letter. The middle branch remembers seeing b in the input, and has not yet counted any letters.

Figure 4 shows the contents of the stack tree after all branches have read the prefix $abb\$bab$. The rightmost branch, has stopped perpetuating the recursion, transitioned to q_e, and emptied its stack. The leftmost branch correctly "guessed" that the a read was the a it was looking for. Subsequently, it transitioned to q_e and removed one # from its stack for the b that was read afterwards. The second branch from the left correctly ignored the first b after the $ sign, and only transitioned to q_e after reading the second b. The second branch from the right is still waiting to find the correct b, and is therefore still is state q_b^2.

3   One-Turn Synchronized Alternating Pushdown Automata

It is a well-known result, due to Ginsburg and Spanier [5], that linear grammars are equivalent to one-turn PDA. One-turn PDA are a sub-family of pushdown automata, where in each computation the stack height switches only once from non-decreasing to non-increasing. That is, once a transition replaces the top symbol of the stack with ϵ, all subsequent transitions may write at most one character. A similar notion of one-turn SAPDA can be defined, where each stack branch can make only one turn in the course of a computation.

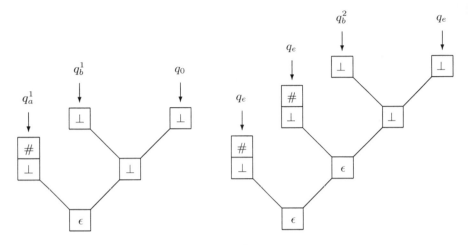

Fig. 3. Stack tree contents after all branches have read ab

Fig. 4. Stack tree contents after all branches have read $abb\$ba$

Definition 10. *Let A be an SAPDA and let T, T' be configurations of A such that $T \vdash T'$.*

- *The computation step (or transition) $T \vdash T'$ is called* increasing, *if it is based on $(q_1, \alpha_1) \wedge \cdots \wedge (q_k, \alpha_k)$ such that $k \geq 2$, or is based on (q, α) such that $|\alpha| \geq 2$.*
- *The computation step (or transition) $T \vdash T'$ is called* decreasing, *if it is based on the collapsing of sibling leaves, or is based on (q, ϵ).*
- *The computation step $T \vdash T'$ is called* non-changing, *if it is based on (q, X), $X \in \Gamma$.*

A computation step is non-decreasing (respectively, non-increasing), if it is non-changing or increasing (respectively, decreasing).

Intuitively, a *turn* is a change from non-decreasing computation steps to non-increasing ones. We would like to capture the number of turns each stack branch makes through the course of a computation. Stack branches are determined by nodes in the configuration tree. We first define the segment of the computation in which a specific node appears, and then we can consider the number of turns for that node.

Definition 11. *Let A be an SAPDA , let $T_0 \vdash \cdots \vdash T_n$ be a computation of A, and let v be a node that appears in a configuration of this computation. Let T_i be the first configuration containing v and let T_j be the last configuration containing v, $0 \leq i \leq j \leq n$. The* span *of node v is the sub-computation beginning with configuration T_i and ending with configuration T_j.*

For example, if v is the root node then the span of v is the entire computation $T_0 \vdash \cdots \vdash T_n$.

Within the span of a node v, not all transitions are relevant to v. For example, transitions may be applied to a node v' which is in a different sub-tree of the configuration. The relevant transitions are, first and foremost, those that are applied to v. However, at some point, v may have child nodes. In this case, transitions applied to v's children are not relevant, but the collapsing of these nodes is. This is because the collapsing of child nodes is similar to popping a symbol from the stack (when one considers the combined stack heights of the parent and the children nodes), and can therefore be viewed as a decreasing transition on v.

Definition 12. *Let A be an SAPDA and let $T_0 \vdash \cdots \vdash T_n$ be a computation of A. Let v be a node that appears in (a configuration of) this computation, and let $T_i \vdash \cdots \vdash T_j$ be its span. Let $t_i, \ldots, t_{j-1}$ be the sequence of transitions applied in the computation $T_i \vdash \cdots \vdash T_j$, i.e., $T_k \vdash T_{k+1}$ by transition t_k, $k = i, \ldots, j-1$. The* relevant transitions *on v is the (order-maintaining) subsequence $t_{\ell_1}, \ldots, t_{\ell_m}$ of $t_i, \ldots, t_{j-1}$ such that for each $k = 1, \ldots, m$, one of the following holds.*

- *The computation step $T_{\ell_k} \vdash T_{\ell_k+1}$ is based on some $(q_1, \alpha_1) \wedge \cdots \wedge (q_k, \alpha_k)$ applied to v, or*
- *the computation step $T_{\ell_k} \vdash T_{\ell_k+1}$ is based on a collapsing of child nodes of v.*

Definition 13. *An SAPDA A is* one-turn, *if for each accepting computation $T_0 \vdash_A \cdots \vdash_A T_n$ the following holds. Let v be a node that appears in (a configuration of) this computation and let $t_{\ell_1}, \ldots, t_{\ell_m}$ be the relevant transitions on v. Then there exists $1 \leq k \leq m$ such that*

- *the computation step $T_{\ell_k} \vdash T_{\ell_k+1}$ is decreasing,*
- *for all $1 \leq k' < k$, the computation steps $T_{\ell_{k'}} \vdash T_{\ell_{k'}+1}$ are non-decreasing, and*
- *for all $k \leq k' < m$, the computation steps $T_{\ell_{k'}} \vdash T_{\ell_{k'}+1}$ are non-increasing.*

Informally, the definition states that in every accepting computation, the hight of each stack branch turns exactly once. Note that the requirement of a decreasing step in the computation is not limiting as we are considering acceptance by empty stack. As such, every accepting computation must have at least one decreasing computation step.

Remark 1. Reordering transitions as necessary, we can assume that all transitions on a node v and its descendants, are applied sequentially.[12] By this assumption, all transitions in the span of a node v are applied either to v or to one of its descendants.

If the automaton is one-turn, these transitions can be partitioned into three parts, the first being non-decreasing transitions applied to v, the second being transitions applied to descendants of v, and the third being non-increasing transitions applied to v. The relevant transitions on v, in this case, are the first and

---

[12] The order of transitions applied to different sub-trees can be changed, without affecting the final configuration reached.

third parts. Note that if the middle section is non-empty then the last transition in the first part must be a conjunctive one which creates child nodes for v (an increasing transition), and the first transition in the final part must be the collapsing of these child nodes (a decreasing transition).

When viewing a classical one-turn PDA as a one-turn SAPDA, there is only one "node" in all configurations, because there are no conjunctive transitions. Therefore, the span of the node is the entire computation, and it is comprised only of the first and third parts, i.e., non-decreasing transitions followed by non-increasing ones, coinciding with the classical notion of a one-turn PDA.

Note that the automaton described in Example 4 is in fact a one-turn SAPDA, while the automaton from Example 3 is not.

4   Linear CG and One-Turn SAPDA

We now show the equivalence of one-turn SAPDA and LCG.

Theorem 1. *A language is generated by an LCG if and only if it is accepted by a one-turn SAPDA.*

The proof of the "only if" part of the theorem is presented in Section 4.1 and the proof of its "if" part is presented in Section 4.2. For full proofs of the theorems and additional details, see [2].

We precede the proof of Theorem 1 with the following immediate corollary.

Corollary 1. *One-turn SAPDA are equivalent to Trellis Automata.*

While both computational models employ a form of parallel processing, their behavior is quite different. To better understand the relationship between the models, see the proof of equivalence between LCG and Trellis Automata in [13]. As LCG and one-turn SAPDA are closely related, the equivalence proof provides intuition on the relation with one-turn SAPDA as well.

4.1   Proof of the "only if" Part of Theorem 1

For the purposes of our proof we assume that grammars do not contain ϵ-rules. Okhotin proved in [10] that it is possible to remove such rules from the grammar, with the exception of an ϵ-rule from the start symbol in the case where ϵ is in the grammar language. We also assume that the start symbol does not appear in the right-hand side of any rule. This can be achieved by augmenting the grammar with a *new* start symbol S', as is done in the classical case.

Let $G = (V, \Sigma, P, S)$ be a linear conjunctive grammar. Consider the SAPDA $A_G = (Q, \Sigma, \Gamma, q_0, \perp, \delta)$, where

- $\Gamma = V \cup \Sigma$ and $\perp = S$,
- $Q = \{q_{u'} :$ for some $A \to (\cdots \& uBy \& \cdots) \in P$ and

$$\text{some } z \in \Sigma^*, \ u = zu'\},^{13}$$

---

[13] That is, u' is a suffix of u.

- $q_0 = q_\epsilon$, and
- δ is defined as follows.
  1. $\delta(q_\epsilon, \epsilon, A)$ is the union of
     (a) $\{(q_\epsilon, w) : A \to w \in P\}$ and
     (b) $\{(q_{u_1}, B_1 y_1) \wedge \cdots \wedge (q_{u_k}, B_k y_k) :$
$$A \to (u_1 B_1 y_1 \ \& \ \cdots \ \& \ u_k B_k y_k) \in P\},$$
  2. $\delta(q_{\sigma u}, \sigma, A) = \{(q_u, A)\}$,
  3. $\delta(q_\epsilon, \sigma, \sigma) = \{(q_\epsilon, \epsilon)\}$

The proof is an extension of the classical one, see, e.g., [6, Theorem 5.3, pp. 115–116]. The construction is modified so as to result in a one-turn automaton. For example, transitions such as transition 2 are used to avoid removing symbols from the stack in the increasing phase of the computation.

Next, we prove that A_G is in fact a one-turn SAPDA. Let $T_0 \vdash \cdots \vdash T_n$ be an accepting computation of A_G, let v be a node appearing in (a configuration of) the computation, and let $T_i \vdash \cdots \vdash T_j$, $0 \leq i \leq j \leq n$, be the span of v.

Proposition 1. *In T_i, the contents of the stack of v is of the form Ay, where $A \in \Gamma$ and $y \in \Sigma^*$.*

Proof. If v is the root, then $T_i = T_0$ and v has S in the stack. If v is not the root, then, since T_i is the first configuration containing v, $T_{i-1} \vdash T_i$ was by a transition of type 1(b). Therefore, v contains Ay in its stack.   □

Proposition 2. *In T_j, the contents of the stack of v is empty.*

Proof. If v is the root, then $T_j = T_n$ and the proposition holds due to the fact that the computation is accepting. If v is not the root, then, since T_i is the last configuration containing v, $T_j \vdash T_{j+1}$ is a collapsing of nodes, one of which is v. Since nodes can only be collapsed with empty stacks, the proposition holds in this case as well.   □

Let $t_{\ell_1}, \ldots, t_{\ell_m}$ be the relevant transitions on v. Corollary 2 below immediately follows from Propositions 1 and 2.

Corollary 2. *There exists $k \in \{1, \ldots, m\}$ such that t_{ℓ_k} is decreasing.*

Now, let k be the smallest index such that t_{ℓ_k} is decreasing (i.e., t_{ℓ_k} is first decreasing transition applied to v). There are three possible cases: t_{ℓ_k} is of type 1(a), where $w = \epsilon$; or it is of type 3; or it is a collapsing of child nodes. In all cases, if the stack is not empty after applying the transition, it contains only terminal symbols. This is because type 1(a) transitions remove the non-terminal symbol, and type 3 and collapsing transitions assume the top symbol is in Σ. By the construction of the automaton, if there is a non-terminal symbol in a stack branch, then it is the top symbol. Therefore, if the top symbol is terminal, so are all the rest. It follows that all subsequent relevant transitions can only be of type 3, meaning they are all decreasing, and A_G is, in fact, one-turn.

4.2   Proof of the "if" Part of Theorem 1

The proof is a variation on a similar proof for the classical case presented in [5]. For the purposes of simplification, we make several assumptions regarding the structure of one-turn SAPDA, namely that they are single-state and that their transitions write at most two symbols at a time. We state that these assumptions are not limiting in the following two lemmas.

Lemma 1. *For each one-turn SAPDA there is an equivalent single-state one-turn SAPDA.*

Definition 14. *Let $A = (\Sigma, \Gamma, \bot, \delta)$ be a single-state SAPDA. We say that A is* bounded *if for all $\sigma \in \Sigma \cup \{\epsilon\}$ and all $X \in \Gamma$ the following holds.*

- *For every $\alpha \in \delta(\sigma, X)$, $|\alpha| \leq 2$.*
- *For every $\alpha_1 \wedge \cdots \wedge \alpha_k \in \delta(\sigma, X)$, $k \geq 2$, $|\alpha_i| = 1$, $i = 1, \ldots, k$.*

Lemma 2. *Every single-state one-turn SAPDA is equivalent to a bounded single-state one-turn SAPDA.*

Let $A = (\Sigma, \Gamma, \bot, \delta)$ be a one-turn SAPDA. By Lemmas 1 and 2, we may assume that A is single-state and bounded. Consider the *linear* conjunctive grammar $G_A = (V, \Sigma, P, S)$, where

- $V = \Gamma \times (\Gamma \cup \{\epsilon\})$,
- $S = [\bot, \epsilon]$, and
- P is the union of the following sets of rules, for all $\sigma \in \Sigma \cup \{\epsilon\}$ and all $X, Y, Z \in \Gamma$.
    1. $\{[X, Y] \rightarrow \sigma[Z, \epsilon] : ZY \in \delta(\sigma, X)\}$,
    2. $\{[X, Y] \rightarrow \sigma[Z, Y] : Z \in \delta(\sigma, X)\}$,
    3. $\{[X, \epsilon] \rightarrow \sigma[Z, \epsilon] : Z \in \delta(\sigma, X)\}$,
    4. $\{[X, \epsilon] \rightarrow (\sigma[X_1, \epsilon] \& \cdots \& \sigma[X_k, \epsilon])$:
$$k \geq 2 \text{ and } X_1 \wedge \cdots \wedge X_k \in \delta(\sigma, X)\},[14]$$
    5. $\{[X, \epsilon] \rightarrow \sigma : \epsilon \in \delta(\sigma, X)\}$,
    6. $\{[X, Y] \rightarrow [X, Z]\sigma : Y \in \delta(\sigma, Z)\}$, and
    7. $\{[X, \epsilon] \rightarrow [X, Z]\sigma : \epsilon \in \delta(\sigma, Z)\}$.

The grammar variables $[X, Y]$ correspond to zero- and one-turn computations starting with X in the stack and ending with Y in the stack. In particular, any word derived from $[\bot, \epsilon]$ is a word with a one-turn emptying computation of A.

The various types of production rules, correspond to the different types of transitions of the automaton.

- Rules of type 1 correspond to increasing computation steps,
- rules of type 2, 3 and 6 correspond to non-changing computation steps,
- rules of type 4 correspond to conjunctive transitions,

---

[14] Since A is bounded, $X_i \neq \epsilon$, $i = 1, \ldots, k$.

- rules of type 5 correspond to the turn step, and
- rules of type 7 correspond to decreasing computation steps.

The correctness of the construction follows from Proposition 3 below, which completes our proof of the "if" part of Theorem 1.

Proposition 3. *For every $X \in \Gamma$, $Y \in \Gamma \cup \{\epsilon\}$, and $w \in \Sigma^*$, $[X, Y] \Rightarrow^* w$ if and only if $(w, X) \vdash^* (\epsilon, Y)$ with exactly one turn.*

5   Mildly Context-Sensitive Languages

Computational linguistics focuses on defining a computational model for natural languages. Originally, context-free languages were considered, and many natural language models are in fact models for context-free languages. However, certain natural language structures that cannot be expressed in context free languages, led to an interest in a slightly wider class of languages which came to be known as *mildly context-sensitive languages* (MCSL). Several formalisms for grammar specification are known to converge to this class [15].

Mildly context sensitive languages are loosely categorized as having the following properties: (1) They contain the context-free languages; (2) They contain such languages as multiple-agreement, cross-agreement and reduplication; (3) They are polynomially parsable; (4) They are semi-linear[15].

There is a strong relation between the class of languages derived by linear conjunctive grammars (and accepted by one-turn SAPDA) and the class of mildly context sensitive languages. The third criterion of MCSL is met by Okhotin's proof that CG membership is polynomial. Multiple-agreement and reduplication with a center marker[16] are shown in Examples 1 and 2 respectively, and an LCG for cross-agreement can be easily constructed. While the first criterion is met for general conjunctive languages, it is not met for the linear sub-family, as there exist context-free languages which are not linear conjunctive [14]. However, every linear context-free language, is of course also a linear conjunctive language.

The fourth criterion of semi-linearity is not met for linear conjunctive languages, and subsequently not for general conjunctive languages. In [14], Okhotin presents an LCG for the language $\{ba^2ba^4 \cdots ba^{2n}b \mid n \in \mathbb{N}\}$, which has super-linear growth. In this respect, LCG and one-turn SAPDA accept some languages not characterized as mildly context-sensitive.

6   Concluding Remarks

We have introduced one-turn SAPDA as a sub-family of SAPDA, and proven that they are equivalent to Linear Conjunctive Grammars. This supports the

---

[15] A language L is *semi-linear* if $\{|w| \mid w \in L\}$ is a finite union of sets of integers of the form $\{l + im \mid i = 0, 1, \dots\}$, $l, m \geq 0$.

[16] Okhotin has conjectured that reduplication without a center marker cannot be generated by any CG. However, this is still an open problem.

claim from [1] that SAPDA are a natural counterpart for CG. The formulation as an automaton provides additional insight into Linear Conjunctive Grammars, and may help solve some of the open questions regarding these grammars.

In [8], Kutrib and Malcher explore a wide range of finite-turn automata with and without turn conditions, and their relationships with closures of linear context-free languages under regular operations. It would prove interesting to explore the general case of finite-turn SAPDA, perhaps finding models for closures of linear conjunctive languages under regular operations.

Acknowledgments

This work was supported by the Winnipeg Research Fund.

References

1. Aizikowitz, T., Kaminski, M.: Conjunctive grammars and alternating pushdown automata. In: Hodges, W., de Queiroz, R. (eds.) WoLLIC 2008. LNCS (LNAI), vol. 5110, pp. 44–55. Springer, Heidelberg (2008)
2. Aizikowitz, T., Kaminski, M.: Linear conjunctive grammars and one-turn synchronized alternating pushdown automata. Technical Report CS-2009-15, Technion – Israel Institute of Technology (2009)
3. Chandra, A.K., Kozen, D.C., Stockmeyer, L.J.: Alternation. Journal of the ACM 28(1), 114–133 (1981)
4. Culik II, K., Gruska, J., Salomaa, A.: Systolic Trellis automata, I and II. International Journal of Computer Mathematics 15 and 16(1 and 3-4), 195–212 and 3–22 (1984)
5. Ginsburg, S., Spanier, E.H.: Finite-turn pushdown automata. SIAM Journal on Control 4(3), 429–453 (1966)
6. Hopcroft, J.E., Ullman, J.D.: Introduction to Automata Theory, Languages and Computation. Addison-Wesley, Reading (1979)
7. Kupferman, O., Vardi, M.Y.: Weak alternating automata are not that weak. ACM Transaction on Computational Logic 2(3), 408–429 (2001)
8. Kutrib, M., Malcher, A.: Finite turns and the regular closure of linear context-free languages. Discrete Applied Mathematics 155(16), 2152–2164 (2007)
9. Ladner, R.E., Lipton, R.J., Stockmeyer, L.J.: Alternating pushdown and stack automata. SIAM Journal on Computing 13(1), 135–155 (1984)
10. Okhotin, A.: Conjunctive grammars. Journal of Automata, Languages and Combinatorics 6(4), 519–535 (2001)
11. Okhotin, A.: A recognition and parsing algorithm for arbitrary conjunctive grammars. Theoretical Computer Science 302, 81–124 (2003)
12. Okhotin, A.: Efficient automaton-based recognition for linear conjunctive languages. International Journal of Foundations of Computer Science 14(6), 1103–1116 (2003)
13. Okhotin, A.: On the equivalence of linear conjunctive grammars and trellis automata. RAIRO Theoretical Informatics and Applications 38(1), 69–88 (2004)
14. Okhotin, A.: On the closure properties of linear conjunctive languages. Theoretical Computer Science 299(1-3), 663–685 (2003)
15. Vijay-Shanker, K., Weir, D.J.: The equivalence of four extensions of context-free grammars. Mathematical Systems Theory 27(6), 511–546 (1994)

A Model-Theoretic Framework for Grammaticality Judgements

Denys Duchier, Jean-Philippe Prost, and Thi-Bich-Hanh Dao

LIFO, Université d'Orléans

Abstract. Although the observation of grammaticality judgements is well acknowledged, their formal representation faces problems of different kinds: linguistic, psycholinguistic, logical, computational. In this paper we focus on addressing some of the logical and computational aspects, relegating the linguistic and psycholinguistic ones in the parameter space. We introduce a model-theoretic interpretation of Property Grammars, which lets us formulate numerical accounts of grammaticality judgements. Such a representation allows for both clear-cut binary judgements, and graded judgements. We discriminate between problems of Intersective Gradience (*i.e.*, concerned with choosing the syntactic category of a model among a set of candidates) and problems of Subsective Gradience (*i.e.*, concerned with estimating the degree of grammatical acceptability of a model). Intersective Gradience is addressed as an optimisation problem, while Subsective Gradience is addressed as an approximation problem.

1 Introduction

Model-Theoretic Syntax (MTS) fundamentally differs from proof-theoretic syntax (or Generative-Enumerative Syntax—GES—as coined by Pullum and Scholz [1]) in the way of representing language: while GES focuses on describing a procedure to generate by enumeration the set of all the legal strings in the language, MTS abstracts away from any specific procedure and focuses on describing individual syntactic properties of language. While the syntactic representation of a string is, in GES, the mere trace of the generative procedure, in MTS it is a model for the grammar, with no information as to how such a model might be obtained. The requirement to be a *model for the grammar* is to satisfy the set of all unordered grammatical constraints.

When compared with GES, the consequences in terms of coverage of linguistic phenomena is significant. Pullum and Scholz have shown that a number of phenomena, which are not accounted for by GES, are well covered in MTS frameworks. Most noticeably, *quasi-expressions*[1] and graded grammaticality judgements are

---

[1] The term *quasi-expression* was coined by Pullum and Scholz [1] in order to refer to those utterances of a natural language, which are not completely well-formed, yet show some form of syntactic structure and properties. In contrast, *expressions* refer to well-formed utterances, that is, utterances which strictly meet all the grammatical requirements. We adopt here the same terminology; we will use *utterance* to refer to either an expression or a quasi-expression.

P. de Groote, M. Egg, and L. Kallmeyer (Eds.): FG 2009, LNAI 5591, pp. 17–30, 2011.
© Springer-Verlag Berlin Heidelberg 2011

only covered by MTS. Yet there exists no logical formulation for such graded grammaticality judgements, although they are made theoretically possible by MTS. This paper proposes such a formulation, based on the model of gradience implemented by Prost [2].

Our contribution is 3-fold: first and foremost, we offer precise model-theoretic semantics for property grammars; we then extend it to permit *loose* models for deviant utterances; and finally we use this formal apparatus to devise scoring functions that can be tuned to agree well with natural comparative judgements of grammaticality.

While Prost [2] proposed a framework for gradience and a parsing algorithm for possibly deviant utterances, his formalization was not entirely satisfactory; among other things, his models were not trees, but technical devices suggested by his algorithmic approach to parsing. Our proposal takes a rather different angle; our models are trees of syntactic categories; our formalization is fully worked out and was designed for easy conversion to constraint programming.

The notions of gradience that underly our approach are described in section 2; property grammars are introduced in section 3; their strong semantics are developed in section 4; their loose semantics in section 5; section 6 presents the postulates that inform our modelization of acceptability judgements, and section 7 provides its quantitative formalization.

2   Gradience

Aarts [3] proposes to discriminate the problems concerned with gradience in two different families: those concerned with *Intersective Gradience* (IG), and those concerned with *Subsective Gradience* (SG). In reference to Set Theory, IG refers to the problem of choosing which category an item belongs to among a set of candidates, while SG refers to the problem of estimating to what extent an item is prototypical within the category it belongs to. Applied here, we regard the choice of a model for an utterance (*i.e.* expression or quasi-expression) as a problem of IG, while the estimation of a degree of grammatical acceptability for a model is regarded as a problem of SG.

For example, Fig 1 illustrates a case of IG with a set of possible parses for a quasi-expression. In that case the preferred model is the first one. The main reason is that, unlike the other ones, it is rooted with the category S.

Fig 2 illustrates different sentences ordered by decreasing grammatical acceptability. Each given judgement corresponds to a (human) estimate of how acceptable it is compared with the reference expression 1.

Fig 3 gives models for quasi-expressions 2 (QE2) and 5 (QE5) from Fig.2. We observe that the model for QE2 is rooted by S, while the one for QE5 is rooted by Star (wildcard category). QE5, unlike QE2, is essentially and crucially missing a VP. QE5 is also unexpectedly terminated with a P. QE2, on the other hand, is only missing a determiner for introducing *rapport*, since it is a requirement in French for a noun to be introduced by a determiner. For all these reasons, the model for QE5 is judged more ungrammatical than the one for QE2.

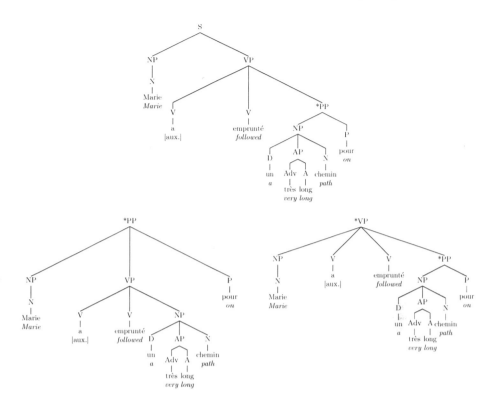

Fig. 1. Intersective Gradience: possible models for the French quasi-expression *Marie a emprunté un très long chemin pour*

1. Les employés ont rendu un rapport très complet à leur employeur [100%]
   The employees have sent a report very complete to their employer
2. Les employés ont rendu rapport très complet à leur employeur [92.5%]
   The employees have sent report very complete to their employer
3. Les employés ont rendu un rapport très complet à [67.5%]
   The employees have sent a report very complete to
4. Les employés un rapport très complet à leur employeur [32.5%]
   The employees a report very complete to their employer
5. Les employés un rapport très complet à [5%]
   The employees a report very complete to their employer

Fig. 2. Sentences of decreasing acceptability

We will come back shortly to the precise meaning of *model*. For the moment, let us just say that a model is a syntactic representation of an utterance. Intuitively, the syntactic representation of an expression can easily be grasped, but it is more problematic in case of a quasi-expression. What we propose in that case, is to *approximate* models, then to choose the optimal one(s). The numeric criterion to be optimised may take different forms ; we decide to maximise the proportion of

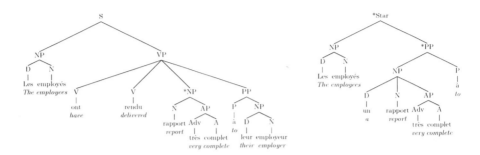

Fig. 3. Models for the quasi-expressions 2. and 5. from Fig.2

grammatical constraints satisfied by the model. Once the problem of IG is solved, we can then make a grammaticality judgement on that model and estimate a degree of acceptability for it. We propose that that estimate be based on different psycholinguistic hypotheses regarding factors of influence in a grammaticality judgement. We propose a formulation for each of them, and for combining them into a single score for the model.

3   Property Grammars

The framework for gradience which we propose is formulated in terms of Property Grammars [4]. Property Grammars are appealing for modeling deviant utterances because they break down the notion of grammaticality into many small constraints (properties) which may be independently violated.

Property Grammars are perhaps best understood as the transposition of phrase structure grammars from the GES perspective into the MTS perspective. Let's consider a phrase structure grammar expressed as a collection of rules. For our purpose, we assume that there is exactly one rule per non-terminal, and that rule bodies may be disjunctive to allow alternate realizations of the same non-terminal. In the GES perspective, such a grammar is interpreted as a generator of strings.

It is important to recognize that the same grammar can be interpreted in the MTS perspective: its models are all the syntax trees whose roots are labeled with the axiom category and such that every rule is satisfied at every node. For example, we say that the rule NP → D N is satisfied at a node if either the node is not labeled with NP, or it has exactly two children, the first one labeled with D and the second one labeled with N.

In this manner, rules have become constraints and a phrase structure grammar can be given model-theoretical semantics by interpretation over syntax tree structures. However these constraints remain very coarse-grained: for example, the rule NP → D N simultaneously stipulates that for a NP, there must be (1) a D child and (2) only one, (3) a N child and (4) only one, (5) nothing else and (6) that the D child must precede the N child.

Property grammars explode rules into such finer-grained constraints called *properties*. They have the form $A : \psi$ meaning in an A, the constraint ψ applies to its children (its constituents). The usual types of properties are:

obligation	$A : \triangle B$	at least one B child
uniqueness	$A : B!$	at most one B child
linearity	$A : B \prec C$	a B child precedes a C child
requirement	$A : B \Rightarrow C$	if there is a B child, then also a C child
exclusion	$A : B \nleftrightarrow C$	B and C children are mutually exclusive
constituency	$A : S?$	the category of any child must be one in S

For the rule NP $\rightarrow$ D N studied above, stipulation (1) would be expressed by a property of *obligation* NP : $\triangle$D, similarly stipulation (3) by NP : $\triangle$N, stipulation (2) by a property of uniqueness NP : D!, similarly stipulation (4) by NP : N!, stipulation (5) by a property of *constituency* NP : $\{$D, N$\}$?, and stipulation (6) by a property of *linearity* NP : D $\prec$ N.

In other publications, property grammars are usually displayed as a collection of boxes of properties. For example, Table 1 contains the property grammar for French that is used in [2]. The present article deviates from the usual presentation in four ways. First, in the interest of brevity, we do not account for features though this would pose no formal problems. Consequently, second: we omit the *dependency* property. Third, we make the constituency property explicit. Fourth, our notation is different: the S box is transcribed as the following set of property literals: S : $\triangle$VP, S : NP!, S : VP!, and S : NP $\prec$ VP.

Table 1. Example property grammar for French

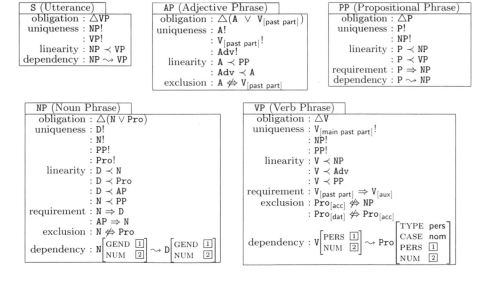

4   Strong Semantics

Property grammars. Let $\mathcal{L}$ be a finite set of labels representing syntactic categories. We write $\mathcal{P}_{\mathcal{L}}$ for the set of all possible property literals over $\mathcal{L}$ formed $\forall c_0, c_1, c_2 \in \mathcal{L}$ in any of the following 6 ways:

$$c_0 : c_1 \prec c_2, \quad c_0 : \triangle c_1, \quad c_0 : c_1!, \quad c_0 : c_1 \Rightarrow c_2, \quad c_0 : c_1 \not\Leftrightarrow c_2, \quad c_0 : s_1?$$

Let $\mathcal{S}$ be a set of elements called *words*. A lexicon is a subset of $\mathcal{L} \times \mathcal{S}$.[2] A *property grammar* G is a pair (P_G, L_G) where P_G is a set of properties (a subset of $\mathcal{P}_{\mathcal{L}}$) and L_G is a lexicon.

Class of models. The strong semantics of property grammars is given by interpretation over the class of syntax tree structures defined below.

We write $\mathbb{N}_0$ for $\mathbb{N} \setminus \{0\}$. A *tree domain* D is a finite subset of $\mathbb{N}_0^*$ which is closed for prefixes and for left-siblings; in other words it satisfies:

$$\forall \pi, \pi' \in \mathbb{N}_0^* \qquad\qquad \pi\pi' \in D \;\Rightarrow\; \pi \in D$$
$$\forall \pi \in \mathbb{N}_0^*, \; \forall i, j \in \mathbb{N}_0 \qquad i < j \wedge \pi j \in D \;\Rightarrow\; \pi i \in D$$

A syntax tree $\tau = (D_\tau, L_\tau, R_\tau)$ consists of a tree domain D_τ, a labeling function $L_\tau : D_\tau \to \mathcal{L}$ assigning a category to each node, and a function $R_\tau : D_\tau \to \mathcal{S}^*$ assigning to each node its surface realization.

For convenience, we define the arity function $A_\tau : D_\tau \to \mathbb{N}$ as follows, $\forall \pi \in D_\tau$:

$$A_\tau(\pi) = \max \{0\} \cup \{i \in \mathbb{N}_0 \mid \pi i \in D_\tau\}$$

Instances. A property grammar G stipulates a set of properties. For example the property $c_0 : c_1 \prec c_2$ is intended to mean that, for a non-leaf node of category c_0, and any two daughters of this node labeled respectively with categories c_1 and c_2, the one labeled with c_1 must precede the one labeled with c_2. Clearly, for each node of category c_0, this property must be checked for every pair of daughters of said node. Thus, we arrive at the notion of instances of a property.

An instance of a property is a pair of a property and a tuple of nodes (paths) to which it is applied. We define the property instances of a grammar G on a syntax tree τ as follows:

$$\mathcal{I}_\tau[\![G]\!] = \cup\{\mathcal{I}_\tau[\![p]\!] \mid \forall p \in P_G\}$$
$$\mathcal{I}_\tau[\![c_0 : c_1 \prec c_2]\!] = \{(c_0 : c_1 \prec c_2)@\langle \pi, \pi i, \pi j \rangle \mid \forall \pi, \pi i, \pi j \in D_\tau, \; i \neq j\}$$
$$\mathcal{I}_\tau[\![c_0 : \triangle c_1]\!] = \{(c_0 : \triangle c_1)@\langle \pi \rangle \mid \forall \pi \in D_\tau\}$$
$$\mathcal{I}_\tau[\![c_0 : c_1!]\!] = \{(c_0 : c_1!)@\langle \pi, \pi i, \pi j \rangle \mid \forall \pi, \pi i, \pi j \in D_\tau, \; i \neq j\}$$
$$\mathcal{I}_\tau[\![c_0 : c_1 \Rightarrow c_2]\!] = \{(c_0 : c_1 \Rightarrow c_2)@\langle \pi, \pi i \rangle \mid \forall \pi, \pi i \in D_\tau\}$$
$$\mathcal{I}_\tau[\![c_0 : c_1 \not\Leftrightarrow c_2]\!] = \{(c_0 : c_1 \not\Leftrightarrow c_2)@\langle \pi, \pi i, \pi j \rangle \mid \forall \pi, \pi i, \pi j \in D_\tau, \; i \neq j\}$$
$$\mathcal{I}_\tau[\![c_0 : s_1?]\!] = \{(c_0 : s_1?)@\langle \pi, \pi i \rangle \mid \forall \pi, \pi i \in D_\tau\}$$

---

[2] We restricted ourselves to the simplest definition sufficient for this presentation.

Pertinence. Since we created instances of all properties in P_G for all nodes in τ, we must distinguish properties which are truly pertinent at a node from those which are not. For this purpose, we define the predicate P_τ over instances as follows:

$$
\begin{aligned}
P_\tau((c_0 : c_1 \prec c_2)@\langle \pi, \pi i, \pi j \rangle) &\equiv L_\tau(\pi) = c_0 \,\wedge\, L_\tau(\pi i) = c_1 \,\wedge\, L_\tau(\pi j) = c_2 \\
P_\tau((c_0 : \triangle c_1)@\langle \pi \rangle) &\equiv L_\tau(\pi) = c_0 \\
P_\tau((c_0 : c_1!)@\langle \pi, \pi i, \pi j \rangle) &\equiv L_\tau(\pi) = c_0 \,\wedge\, L_\tau(\pi i) = c_1 \,\wedge\, L_\tau(\pi j) = c_1 \\
P_\tau((c_0 : c_1 \Rightarrow c_2)@\langle \pi, \pi i \rangle) &\equiv L_\tau(\pi) = c_0 \,\wedge\, L_\tau(\pi i) = c_1 \\
P_\tau((c_0 : c_1 \not\Leftrightarrow c_2)@\langle \pi, \pi i, \pi j \rangle) &\equiv L_\tau(\pi) = c_0 \,\wedge\, (L_\tau(\pi i) = c_1 \,\vee\, L_\tau(\pi j) = c_2) \\
P_\tau((c_0 : s_1?)@\langle \pi, \pi i \rangle) &\equiv L_\tau(\pi) = c_0
\end{aligned}
$$

Satisfaction. When an instance is pertinent, it should also (preferably) be satisfied. For this purpose, we define the predicate S_τ over instances as follows:

$$
\begin{aligned}
S_\tau((c_0 : c_1 \prec c_2)@\langle \pi, \pi i, \pi j \rangle) &\equiv i < j \\
S_\tau((c_0 : \triangle c_1)@\langle \pi \rangle) &\equiv \vee\{ L_\tau(\pi i) = c_1 \mid 1 \le i \le A_\tau(\pi) \} \\
S_\tau((c_0 : c_1!)@\langle \pi, \pi i, \pi j \rangle) &\equiv i = j \\
S_\tau((c_0 : c_1 \Rightarrow c_2)@\langle \pi, \pi i \rangle) &\equiv \vee\{ L_\tau(\pi j) = c_2 \mid 1 \le j \le A_\tau(\pi) \} \\
S_\tau((c_0 : c_1 \not\Leftrightarrow c_2)@\langle \pi, \pi i, \pi j \rangle) &\equiv L_\tau(\pi i) \ne c_1 \,\vee\, L_\tau(\pi j) \ne c_2 \\
S_\tau((c_0 : s_1?)@\langle \pi, \pi i \rangle) &\equiv L_\tau(\pi i) \in s_1
\end{aligned}
$$

We write $I^0_{G,\tau}$ for the set of pertinent instances of G in τ, $I^+_{G,\tau}$ for its subset that is satisfied, and $I^-_{G,\tau}$ for its subset that is violated:

$$
\begin{aligned}
I^0_{G,\tau} &= \{ r \in \mathcal{I}_\tau[\![G]\!] \mid P_\tau(r) \} \\
I^+_{G,\tau} &= \{ r \in I^0_{G,\tau} \mid S_\tau(r) \} \\
I^-_{G,\tau} &= \{ r \in I^0_{G,\tau} \mid \neg S_\tau(r) \}
\end{aligned}
$$

Admissibility. A syntax tree τ is admissible as a candidate model for grammar G iff it satisfies the projection property,[3] i.e. $\forall \pi \in D_\tau$:

$$
A_\tau(\pi) = 0 \quad \Rightarrow \quad \langle L_\tau(\pi), R_\tau(\pi) \rangle \in L_G
$$

$$
A_\tau(\pi) \ne 0 \quad \Rightarrow \quad R_\tau(\pi) = \sum_{i=1}^{i = A_\tau(\pi)} R_\tau(\pi i)
$$

where $\sum$ represents here the concatenation of sequences. In other words: leaf nodes must conform to the lexicon, and interior nodes pass upward the ordered realizations of their daughters. We write $\mathcal{A}_G$ for the set of admissible syntax trees for grammar G.

---

[3] It should be noted that Prost [2] additionally requires that all constituency properties be satisfied.

Strong models. A syntax tree τ is a strong model of a property grammar G iff it is admissible and $I_{G,\tau}^{-} = \emptyset$. We write $\tau : \sigma \models G$ iff τ is a strong model of G with realization σ, i.e. such that $R_\tau(\varepsilon) = \sigma$.

5   Loose Semantics

Since property grammars are intended to also account for deviant utterances, we must define alternate semantics that accommodate deviations from the *strong* interpretation. The *loose semantics* will allow some property instances to be violated, but will seek syntax trees which maximize the overall *fitness* for a specific utterance.

Admissibility. A syntax tree τ is loosely admissible for utterance σ iff it is admissible and its realization is $\sigma = R_\tau(\epsilon)$. We write $\mathcal{A}_{G,\sigma}$ for the loosely admissible syntax trees for utterance σ:

$$\mathcal{A}_{G,\sigma} = \{\tau \in \mathcal{A}_G \mid R_\tau(\epsilon) = \sigma\}$$

Following Prost [2], we define fitness as the ratio of satisfied pertinent instances over the total number of pertinent instances:

$$F_{G,\tau} = I_{G,\tau}^{+}/I_{G,\tau}^{0}$$

The loose models for an utterance σ are all loosely admissible models for utterance σ that maximize fitness:

$$\tau : \sigma \approx G \quad \text{iff} \quad \tau \in \operatorname*{argmax}_{\tau' \in \mathcal{A}_{G,\sigma}}(F_{G,\tau'})$$

6   Modeling Judgements of Acceptability

We now turn to the issue of modeling natural judgements of acceptability. We hypothesize that an estimate for acceptability can be predicted by quantitative factors derivable from the loose model of an utterance. To that end, we must decide what factors and how to combine them. The answers we propose are informed by the 5 postulates outlined below.

Postulates 1, 2, and 3 are substantiated by empirical evidence and work in the fields of Linguistics and Psycholinguistics, but postulates 4 and 5 are speculative. While the factors we consider here are all syntactic in nature, it is clear that a complete model for human judgements of acceptability should also draw on other dimensions of language (semantics, pragmatics, . . .).

Postulate 1 (Failure Cumulativity). *Gradience is impacted by constraint failures; that is, an utterance's acceptability is impacted by the number of constraints it violates.*

This factor is probably the most intuitive, and is the most commonly found in the literature [5, 6]. It corresponds to Keller's *cumulativity effect*, substantiated by empirical evidence.

Postulate 2 (Success Cumulativity). *Gradience is impacted by constraint successes; that is, an utterance acceptability is impacted by the number of constraints it satisfies.*

Different works suggest that acceptability judgements can also be affected by successful constraints [7, 8, 9, 3]. The underlying intuition is that failures alone are not sufficient to account for acceptability, hence the postulate that some form of interaction between satisfied and violated constraints contributes to the judgements. It significantly differs from other accounts of syntactic gradience, which only rely on constraint failures (*e.g.* Keller's LOT, or Schröder's WCDG).

Postulate 3 (Constraint Weighting). *Acceptability is impacted to a different extent according to which constraint is satisfied or violated.*

Here we postulate that constraints are weighted according to their influence on acceptability. This intuition is commonly shared in the literature[4] [8, 9, 10, 11, 12, 13, 5, 14, 15], and supported by empirical evidence. In this paper, we make the simplifying assumption that a constraint weighs the same whether it is satisfied or violated, but we could just as well accommodate different weights.

Strategies for assigning weights may be guided by different considerations of *scope* and *granularity*. Scope: should weights be assigned to constraint individually or by constraint type. Granularity: should weights be decided globally for the grammar, or separately for each syntactic category.

Scope and granularity can then be combined in different ways: all constraints of the same type at the grammar level, or all constraints of the same type at the construction level, or individual constraints at the construction level, or individual constraints at the grammar level. The difference between the last two possibilities assumes that the same constraint may occur in the specification of more than one construction.

Arguably, the finer the granularity and the narrower the scope, the more flexibility and accuracy we get but at a significant cost in maintenance. This high cost is confirmed by Schröder [6], who opted for weights being assigned to each individual constraint at the grammar level. Prost [2] opted for a compromise, where the weighting scheme is restricted to the constraint types at the grammar level, which means that all constraints of the same type in the grammar are assigned the same weight. For example, all the Linearity constraints (*i.e.* word order) are weighted 20, all the Obligation constraints (*i.e.* heads) are weighted 10, and so on.

While the strategy of weight assignment is of considerable methodological import, for the purpose of the present logical formulation it is sufficient to suppose given a function that maps each constraint to its weight.

Postulate 4 (Constructional complexity). *Acceptability is impacted by the complexity of the constituent structure.*

---

[4] Note that these constraint weights take different meanings in different works. Many of them, for instance, have a statistical component that we do not have here.

How to precisely measure and capture the complexity of an utterance is an open question, which we do not claim to fully answer. In fact, this factor of influence probably ought to be investigated in itself, and split into more fine-grained postulates with respect to acceptability and syntactic gradience. Different works from Gibson [16, 12] could be used as a starting point for new postulates in this regard. Here we simply measure the complexity of the category a constituent belongs to as the number of constraints specifying this category in the grammar. This postulate aims to address, among others, the risk of disproportionate convergence. The underlying idea is to balance the number of violations with the number of specified constraints: without such a precaution a violation in a rather simple construction, such as AP, would be proportionally much more costly than a violation in a rather complex construction, such as NP.

Postulate 5 (Propagation). *Acceptability is propagated through the dominance relationships; that is, an utterance's acceptability depends on its nested constituents' acceptability.*

Here it is simply postulated that the nested constituents' acceptability is recursively propagated to their dominant constituent.

7   Formalizing Judgements of Acceptability

Following the previous discussion, we define a *weighted property grammar* G as a triple (P_G, L_G, ω_G) where (P_G, L_G) is a property grammar and $\omega_G : P_G \to \mathbb{R}$ is a function assigning a weight to each property.

Since our approach relies on quantitative measurements of satisfactions and violations, it must be formulated in terms of property instances. Furthermore, postulates 4 and 5 require the computation of local quantitative factors at each node in the model. For this reason, we need to identify, in the set of all property instances the subset which applies at a given node.

For each property instance r, we write $\mathsf{at}(r)$ for the node where it applies; and we define it by cases $\forall p \in \mathcal{P}_{\mathcal{L}}$, $\forall \pi_0, \pi_1, \pi_2 \in \mathbb{N}_0^*$ as follows:

$$\mathsf{at}(p@\langle \pi_0 \rangle) = \pi_0 \qquad \mathsf{at}(p@\langle \pi_0, \pi_1 \rangle) = \pi_0 \qquad \mathsf{at}(p@\langle \pi_0, \pi_1, \pi_2 \rangle) = \pi_0$$

If B is a set of instances, then $B|_\pi$ is the subset of B of all instances applying at node π:

$$B|_\pi = \{ r \in B \mid \mathsf{at}(r) = \pi \}$$

We now define the sets of instances pertinent, satisfied, and violated at node π:

$$I_{G,\tau,\pi}^0 = I_{G,\tau}^0|_\pi \qquad\qquad I_{G,\tau,\pi}^+ = I_{G,\tau}^+|_\pi \qquad\qquad I_{G,\tau,\pi}^- = I_{G,\tau}^-|_\pi$$

which allow us to express the cumulative weights of pertinent, satisfied, and violated instances at node π:

$$W^0_{G,\tau,\pi} = \sum \{\omega_G(x) \mid \forall x @ y \in I^0_{G,\tau,\pi}\}$$

$$W^+_{G,\tau,\pi} = \sum \{\omega_G(x) \mid \forall x @ y \in I^+_{G,\tau,\pi}\}$$

$$W^-_{G,\tau,\pi} = \sum \{\omega_G(x) \mid \forall x @ y \in I^-_{G,\tau,\pi}\}$$

Following Prost [2], we define at each node π the quality index $W_{G,\tau,\pi}$, the satisfaction ratio $\rho^+_{G,\tau,\pi}$, and the violation ratio $\rho^-_{G,\tau,\pi}$:

$$W_{G,\tau,\pi} = \frac{W^+_{G,\tau,\pi} - W^-_{G,\tau,\pi}}{W^+_{G,\tau,\pi} + W^-_{G,\tau,\pi}} \qquad \rho^+_{G,\tau,\pi} = \frac{|I^+_{G,\tau,\pi}|}{|I^0_{G,\tau,\pi}|} \qquad \rho^-_{G,\tau,\pi} = \frac{|I^-_{G,\tau,\pi}|}{|I^0_{G,\tau,\pi}|}$$

According to postulate 4, we must take into account the complexity of a construction: is it specified by many properties or by few? For each node π, we look up the set of properties $T_{G,\tau,\pi}$ that are used in the grammar to specify the category $L_\tau(\pi)$ of π:

$$T_{G,\tau,\pi} = \{c : C \in P_G \mid L_\tau(\pi) = c\}$$

and we use it to define a completeness index $C_{G,\tau,\pi}$:

$$C_{G,\tau,\pi} = \frac{|I^0_{G,\tau,\pi}|}{|T_{G,\tau,\pi}|}$$

According to postulate 5, these quantities must be combined recursively to compute the overall rating of a model. Several rating functions have been investigated: we describe the *index of grammaticality* and the *index of coherence*.

7.1   Index of Grammaticality

This scoring function is based on a local compound factor called the *index of precision* computed as follows:

$$P_{G,\tau,\pi} = kW_{G,\tau,\pi} + l\rho^+_{G,\tau,\pi} + mC_{G,\tau,\pi}$$

where the parameters (k, l, m) are used to tune the model. The index of grammaticality at node π is then defined inductively thus:

$$g_{G,\tau,\pi} = \begin{cases} P_{G,\tau,\pi} \cdot \frac{1}{A_\tau(\pi)} \sum_{i=1}^{A_\tau(\pi)} g_{G,\tau,\pi i} & \text{if } A_\tau(\pi) \neq 0 \\ 1 & \text{if } A_\tau(\pi) = 0 \end{cases}$$

The overall score of a loose model τ is the score $g_{G,\tau,\varepsilon}$ of its root node.

7.2   Index of Coherence

This scoring function is based on a local compound factor called the *index of anti-precision* computed as follows:

$$A_{G,\tau,\pi} = kW_{G,\tau,\pi} - l\rho^-_{G,\tau,\pi} + mC_{G,\tau,\pi}$$

where the parameters (k, l, m) are used to tune the model. The index of coherence at node π is then defined inductively thus:

$$\gamma_{G,\tau,\pi} = \begin{cases} A_{G,\tau,\pi} \cdot \frac{1}{A_\tau(\pi)} \sum_{i=1}^{A_\tau(\pi)} \gamma_{G,\tau,\pi i} & \text{if } A_\tau(\pi) \neq 0 \\ 1 & \text{if } A_\tau(\pi) = 0 \end{cases}$$

The overall score of a loose model τ is the score $\gamma_{G,\tau,\varepsilon}$ of its root node.

7.3 Experimental Validation and Perspectives

An interesting aspect of the framework presented here is that it makes it possible to formally devise scoring functions such as those for the indexes of Grammaticality or Coherence. It is interesting because it opens the door to reasonning with graded grammaticality, in relying on numerical tools which can be validated empirically. As far as Grammaticality and Coherence are concerned, they find their origin in psycholinguistic postulates, but other kinds of justifications may just as well yield different formulations.

Of course, assigning a score to an utterance is only meaningful if supporting evidence can be put forward for validation. In the present case, the relevance of these automatic scores has been validated experimentally in Blache *et al.* [9] then in Prost [2], in measuring to what extent the scores correlate to human judgements. Human judgements of acceptability were gathered for both grammatical and ungrammatical sentences, as part of psycholinguistic experiments (reported in [9]) using the Magnitude Estimation protocol [17]. The corpus in use was artificially constructed, each sentence matching one of the 20 different error patterns. The experiment involves 44 annotators, all native speakers of French, and with no particular knowledge of linguistics. Each annotator was asked to rate, for each sentence, how much better (or worse) it was compared with a reference sentence. The figures obtained were normalised across annotators for every sentence, providing a score of human judgement for each of them.

The validation of the automatic scoring was then performed in calculating the correlation between the mean scores (automatic judgements on one hand, and human judgements on the other hand) per error pattern. The outcome is a Pearson's correlation coefficient $\rho = 0.5425$ for the Coherence score, and $\rho = 0.4857$ for the Grammaticality Index.

Interestingly as well, the same scoring functions are also opened to the modelling of graded judgements of grammaticality, as opposed to ungrammaticality. The scale of scores calculated through the Grammaticality or Coherence Indexes being open-ended, two distinct expressions can be assigned two distinct scores. We could then wonder whether, and to what extent, these scores are comparable. If comparable, then they could be interpreted as modelling the linguistic complexity of an expression. For example, we might like scores to capture that an expression with embedded relative clauses is more complex to comprehend than a simple Subject-Verb-Object construction. Unfortunately, the psycholinguistic experiment in [9] was not designed to that end (very few different constructions are in fact used for the reference sentences). Validating that hypothesis is, therefore, not possible at this stage, and must be kept for further works.

Finally, our formalization was designed for easy conversion to constraint programming and we are currently developing a prototype solver to compare its practical performance with the baseline established by Prost's parser.

8   Conclusion

While formal grammars typically limit their scope to well-formed utterances, we wish to extend their formal reach into the area of graded grammaticality. The first goal is to permit the analysis of well-formed and ill-formed utterances alike. The second more ambitious goal is to devise accurate models of natural judgements of acceptability.

In this paper, we have shown that property grammars are uniquely suited for that purpose. We contributed precise model-theoretic semantics for property grammars. Then, we relaxed these semantics to permit the loose models required by deviant utterances. Finally, we showed how formally to devise quantitative scoring functions on loose models.

Prost [2] has shown that these scoring functions can be tuned to agree well with human judgements.

Acknowledgement

This work was partly funded by the French *Agence Nationale de la Recherche*, project ANR-07-MDCO-03 (CRoTAL).

References

1. Pullum, G., Scholz, B.: On the distinction between model-theoretic and generative-enumerative syntactic frameworks. In: de Groote, P., Morrill, G., Rétoré, C. (eds.) LACL 2001. LNCS (LNAI), vol. 2099, pp. 17–43. Springer, Heidelberg (2001)
2. Prost, J.P.: Modelling Syntactic Gradience with Loose Constraint-based Parsing. Cotutelle Ph.D. Thesis, Macquarie University, Sydney, Australia, and Université de Provence, Aix-en-Provence, France (December 2008)
3. Aarts, B.: Syntactic gradience: the nature of grammatical indeterminacy. Oxford University Press, Oxford (2007)
4. Blache, P.: Constraints Graphs and Linguistic Theory (source unknown) (2000)
5. Keller, F.: Gradience in Grammar - Experimental and Computational Aspects of Degrees of Grammaticality. PhD thesis, University of Edinburgh (2000)
6. Schröder, I.: Natural Language Parsing with Graded Constraints. PhD thesis, Universität Hamburg (2002)
7. Aarts, B.: Modelling Linguistic Gradience. Studies in Language 28(1), 1–49 (2004)
8. Blache, P., Prost, J.P.: Gradience, constructions and constraint systems. In: Christiansen, H., Skadhauge, P.R., Villadsen, J. (eds.) CSLP 2005. LNCS (LNAI), vol. 3438, pp. 74–89. Springer, Heidelberg (2005) (revised and extended version)
9. Blache, P., Hemforth, B., Rauzy, S.: Acceptability prediction by means of grammaticality quantification. In: Proceedings of the 21st International Conference on Computational Linguistics and 44th Annual Meeting of the Association for Computational Linguistics, Sydney, Australia, pp. 57–64. Association for Computational Linguistics (July 2006)

10. Bresnan, J., Nikitina, T.: On the Gradience of the Dative Alternation. Draft (May 7, 2003)

11. Foth, K.: Writing Weighted Constraints for Large Dependency Grammars. Proceedings of Recent Advances in Dependency Grammars, COLING-Workshop (2004)

12. Gibson, E.: The Dependency Locality Theory: A Distance-Based Theory of Linguistic Complexity. In: Marantz, A., Miyashita, Y., O'Neil, W. (eds.) Image, Language, Brain, pp. 95–126. MIT Press, Cambridge (2000)

13. Heinecke, J., Kunze, J., Menzel, W., Shröder, I.: Eliminative Parsing with Graded Constraints. In: Proceedings 7th International Conferenceon Computational Linguistics, 36th Annual Meeting of the ACL, Coling–ACL 1998, Montreal, Canada, pp. 526–530 (1998)

14. Sorace, A., Keller, F.: Gradience in linguistic data. Lingua 115(11), 1497–1524 (2005)

15. VanRullen, T.: Vers une analyse syntaxique à granularité variable. PhD thesis, Université de Provence, Informatique (2005)

16. Gibson, E., Schütze, C.T., Salomon, A.: The relationship between the frequency and the processing complexity of linguistic structure. Journal of Psycholinguistic Research 25(1), 59–92 (1996)

17. Bard, E., Robertson, D., Sorace, A.: Magnitude Estimation of Linguistic Acceptability. Language 72(1), 32–68 (1996)

Multi-Component Tree Insertion Grammars

Pierre Boullier and Benoît Sagot

Alpage, INRIA Paris-Rocquencourt & Université Paris 7
Domaine de Voluceau — Rocquencourt, BP 105
78153 Le Chesnay Cedex, France
{Pierre.Boullier,Benoit.Sagot}@inria.fr

Abstract. In this paper we introduce a new mildly context sensitive formalism called Multi-Component Tree Insertion Grammar. This formalism is a generalization of Tree Insertion Grammars in the same sense that Multi-Component Tree Adjoining Grammars is a generalization of Tree Adjoining Grammars. We show that this class of grammatical formalisms is equivalent to Multi-Component Tree Adjoining Grammars, and that it also defines a hierarchy of languages whose supplementary formal power between two increasing levels is more gently delivered than the one given by Multi-Component Tree Adjoining Grammars. We show that Multi-Component Tree Insertion Grammars and simple Range Concatenation Grammars are equivalent and we show how to transform a grammar of one type into an equivalent grammar of the other type. Such a transformation gives a method to build efficient parsers for Multi-Component Tree Insertion Languages.

1 Introduction

The notion of mild context-sensitivity (MCS) [1,2] is an attempt to express the formal power needed to define the syntax of natural languages. However, all incarnations of MCS formalisms are not equivalent. On the one hand, near the bottom of the hierarchy, we find tree adjoining grammars (TAGs) [3], the most popular mild context-sensitive formalism, and some other weakly equivalent formalisms. On the other hand, near the top of the hierarchy, we find Multi-Component Tree Adjoining Grammars (MCTAGs) which are, in turn and among others, equivalent to Linear Context-Free Rewriting Systems (LCFRSs) [4].

In this paper we introduce a new mild context-sensitive formalism, the multi-component tree insertion grammar (MCTIG) which plays, w.r.t. tree insertion grammar TIG the same role as MCTAG plays w.r.t. TAG. We know that TIGs are weakly equivalent to context-free grammars (CFGs) but they allow to build (parse) structures that are not possible to build with CFGs. From a practical point of view we know that TIGs, which may be seen as a restriction of TAGs, are of importance since many usual linguistic constructions do not need the power of TAGs. On the other hand there exist linguistic constructions that cannot be expressed with TAGs. Thus MCTAGs have been introduced. These various formalisms named k-MCTAGs form a strict hierarchy which depends upon an

P. de Groote, M. Egg, and L. Kallmeyer (Eds.): FG 2009, LNAI 5591, pp. 31–46, 2011.

integer k. TAGs are 1-MCTAGs and the languages of k-MCTAGs (k-MCTALs) are strictly included into $k + 1$-MCTALs. However the extra power earned when we use $k + 1$-MCTAGs instead of k-MCTAGs is too large. For example let L_i be the counting language for i ($L_i = \{a_1^n \ldots a_i^n \mid n \geq 0\}$). It is well known that 1-MCTAGs (i.e., TAGs) can define L_1, L_2, L_3 and L_4 but not L_5. If we want to define L_5 we must use 2-MCTAGs. However, doing that we can also define L_6, L_7 and L_8. We shall see that the power of k-MCTIGs is more balanced.

On the other hand, though being non MCS, there is a very attractive powerful formalism named Range Concatenation Grammar (RCG). Its power comes from the fact that it exactly covers the class PTIME and its attractivity comes from both theoretical and practical considerations. From a theoretical point of view it possesses many closure properties among which the closure by intersection is the most salient. From a practical point of view there exist very efficient polynomial parse time parsers. For example TAGs and MCTAGs have been transformed into equivalent RCGs [5,6] whose parsers achieved top-of-the-art efficiency [7].

In this paper we define MCTIGs and study its relationships with both MC-TAGs and RCGs, thus exhibiting a way to produce an efficient MCTIG parser.

2   Basic Notions and Notations

2.1   Tree Insertion Grammars

We suppose that the reader is familiar with Tree Adjoining Grammars (TAGs), as defined in, e.g., [3].

Tree Insertion Grammar (TIG) [8] is a variant of TAG in which the auxiliary tree shapes have a restricted form: wrapping auxiliary trees are prohibited leaving only left and right auxiliary trees.[1]

A *left auxiliary tree* (resp. *right auxiliary tree*) is an auxiliary tree in which the foot node is the rightmost (resp. leftmost) leaf node and in which adjunction nodes of its spine only select left auxiliary trees (resp. right auxiliary trees) or empty auxiliary

Left Auxiliary Tree

Right Auxiliary Tree

Empty Auxiliary Tree

Thus the set of auxiliary trees $\mathcal{A}$ can be partitioned into three sets $\mathcal{L}$, $\mathcal{R}$ and $\mathcal{E}$, respectively the set of left, right and empty trees.

It is a well known result [8] that TIGs are (weakly) equivalent to CFGs and can be parsed in $\mathcal{O}(n^3)$ time.

---

[1] In the original definition of TIGs, there exist other differences which are not considered in this article. In particular we disallow simultaneous adjunctions at a single node (they can be performed by the usual adjunction operation), but we allow empty auxiliary trees.

2.2   Multi-Component Tree Adjoining Grammars

Multi-component TAGs (MCTAGs) are an extension of TAGs which was introduced in [9] and later refined in [3], in which the adjunction operation involves a set of auxiliary trees instead of a single auxiliary tree. In a MCTAG, the elementary structures, both initial and auxiliary, instead of being two sets of single trees, consist of two finite sets of (ordered) finite tree sets. In MCTAGs, the adjunction operation of an auxiliary tree set is defined as the simultaneous adjunction of each of its component trees and accounts for a single step in the derivation process. This multi-component adjunction (MCA) operation is defined as follows. All the trees of an auxiliary tree set can be adjoined into distinct nodes (addresses) in a single elementary (initial or auxiliary) tree set. If the maximum cardinality of the elementary tree sets is k, we have a k-MCTAG. Of course, if the cardinality of each tree set is one, a 1-MCTAG is a TAG. In [2], the author has shown that the languages defined by MCTAGs are equal to the languages defined by Linear Context-Free Rewriting Systems (LCFRSs) [4].

We assume that (multi-component) substitution operations are disallowed and are replaced by MCA operations. Thus, without loss of generality, we assume that initial tree sets are singletons whose root nodes are all labelled by the start symbol S. Moreover, we assume that adjunction is allowed neither at the root nor at the foot of any tree and that MCAs are mandatory on inside nonterminal nodes which are thus called *adjunction nodes*.

We can think of two types of *locality* for MCAs, one type, named *tree locality*, requires that all trees in an auxiliary tree set adjoin to a unique tree of an elementary tree set; the other type, named *set locality*, requires that all trees in an auxiliary tree set adjoin to the same elementary tree set, not necessarily to a unique tree and not necessarily to all the trees in this elementary tree set. We choose to consider the set-local interpretation of the term since it is more general than the tree-local version and is the one equivalent to LCFRS.

Example 1. For example, let G be a 3-MCTAG in which the set of initial tree sets is $\mathcal{I} = \{\{\alpha\}\}$ and the set of auxiliary tree sets is $\mathcal{A} = \{\beta_1, \beta_2, \beta_3\}$, where each tree set β_1, β_2 and β_3 contains three trees: $\beta_1 = \{\beta_{11}, \beta_{12}, \beta_{13}\}$, $\beta_2 = \{\beta_{21}, \beta_{22}, \beta_{23}\}$ and $\beta_3 = \{\beta_{31}, \beta_{32}, \beta_{33}\}$. Its elementary trees are depicted below.[2]

We can easily see that the language defined by G is the non-TAL three-copy language $\{www \mid w \in \{a, b\}^*\}$. Each simultaneous adjunction of an auxiliary tree set β_1 or β_2 in the elementary tree sets α, β_1 or β_2 produces respectively three related a's or three related b's, while the MCA of (the trees in) β_3 terminates the process.

---

[2] The same denotation is used both for nonterminal nodes and their addresses, while terminal nodes are denoted by their terminal labels or ε. The root node of an elementary tree τ is also denoted by τ. Nonterminal nodes are annotated by their nonterminal labels while terminal nodes have no annotations. In auxiliary trees, foot node labels are marked by an $*$. For example, the root of tree β_{11} is the node β_{11}, whose label is A, and the foot node of this tree is $\beta_{11}.2.1$ and its label is A^*.

$$\alpha : \left\{ \begin{matrix} & \alpha\ S & \\ \alpha.1\ A & \alpha.2\ B & \alpha.3\ C \\ \varepsilon & \varepsilon & \varepsilon \end{matrix} \right\} \quad \beta_1 : \left\{ \begin{matrix} & \beta_{11}\ A & & \beta_{12}\ B & & \beta_{13}\ C \\ a & \beta_{11}.2\ A & a & \beta_{12}.2\ B & a & \beta_{13}.2\ C \\ & \beta_{11}.2.1\ A^* & & \beta_{12}.2.1\ B^* & & \beta_{13}.2.1\ C^* \end{matrix} \right\}$$

$$\beta_3 : \left\{ \begin{matrix} \beta_{31}\ A & \beta_{32}\ B & \beta_{33}\ C \\ \beta_{31}.2\ A^* & \beta_{32}.2\ B^* & \beta_{33}.2\ C^* \end{matrix} \right\} \quad \beta_2 : \left\{ \begin{matrix} & \beta_{21}\ A & & \beta_{22}\ B & & \beta_{23}\ C \\ b & \beta_{21}.2\ A & b & \beta_{22}.2\ B & b & \beta_{23}.2\ C \\ & \beta_{21}.2.1\ A^* & & \beta_{22}.2.1\ B^* & & \beta_{23}.2.1\ C^* \end{matrix} \right\}$$

2.3 Positive Range Concatenation Grammars

A *positive range concatenation grammar* (PRCG) $G = (N, T, V, P, S)$ is a 5-tuple in which:

- T and V are disjoint alphabets of *terminal symbols* and *variable symbols* respectively.
- N is a non-empty finite set of *predicates* of fixed *arity* (also called *fan-out*). We write $k = arity(A)$ if the arity of the predicate A is k. A predicate A and its arguments is noted $A(\boldsymbol{\alpha})$ with a vector notation s.t. $|\boldsymbol{\alpha}| = k$ and $\boldsymbol{\alpha}[j]$ is its jth argument. An argument is a string in $(V \cup T)^*$.
- S is a distinguished predicate called the *start predicate* (or *axiom*) of arity 1.
- P is a finite set of *clauses*. A clause c is a rewriting rule of the form $A_0(\boldsymbol{\alpha_0}) \to A_1(\boldsymbol{\alpha_1}) \ldots A_r(\boldsymbol{\alpha_r})$ where r, $r \geq 0$ is its *rank*. By definition $c[i] = A_i(\boldsymbol{\alpha_i})$, $0 \leq i \leq r$ where A_i is a predicate together with $\boldsymbol{\alpha_i}$ its arguments and $c[i][j]$ is its jth argument $X_1 \ldots X_{n_{ij}}$ (the X_k's are terminal or variable symbols), while $c[i][j][k]$, $0 \leq k \leq n_{ij}$ is a *position*.

For a given clause c, each *subargument occurrence* is denoted by a pair of positions $(c[i][j][k], c[i][j][k'])$ with $k \leq k'$.

Let $w = a_1 \ldots a_n$ be an input string in T^*, each occurrence of a substring $a_{l+1} \ldots a_u$ is a pair of positions $(w[l], w[u])$ s.t. $0 \leq l \leq u \leq n$ called a *range* and noted $\langle l..u \rangle_w$ or $\langle l..u \rangle$ when w is implicit. In the range $\langle l..u \rangle$, l is its *lower bound* while u is its *upper bound*. If $l = u$, the range $\langle l..u \rangle$ is an *empty* range, it spans an empty substring. If $\rho_1 = \langle l_1..u_1 \rangle$, $\ldots$ and $\rho_m = \langle l_m..u_m \rangle$ are ranges, the *concatenation* of $\rho_1, \ldots, \rho_m$ noted $\rho_1 \ldots \rho_m$ is the range $\rho = \langle l..u \rangle$ if and only if we have $u_i = l_{i+1}$, $1 \leq i < m$, $l = l_1$ and $u = u_m$.

If $c = A_0(\boldsymbol{\alpha_0}) \to A_1(\boldsymbol{\alpha_1}) \ldots A_r(\boldsymbol{\alpha_r})$ is a clause, each of its subargument occurrence $(c[i][j][k], c[i][j][k'])$ may take a range $\rho = \langle l..u \rangle$ as value, in that case, we say that it is *instantiated* by ρ.

- If the subargument is the empty string (i.e., $k = k'$), ρ is an empty range.
- If the subargument is a terminal symbol (i.e., $k + 1 = k'$ and $X_{k'} \in T$), ρ is such that $l + 1 = u$ and $a_u = X_{k'}$. Note that several occurrences of the same terminal symbol may be instantiated by different ranges.
- If the subargument is a variable symbol (i.e., $k + 1 = k'$ and $X_{k'} \in V$), any occurrence $(c[i'][j'][m], c[i'][j'][m'])$ of $X_{k'}$ is instantiated by ρ. Thus, each occurrence of the same variable symbol must be instantiated by the same range.

- If the subargument is the string $X_{k+1} \ldots X_{k'}$, ρ is its instantiation if and only if we have $\rho = \rho_{k+1} \ldots \rho_{k'}$ in which $\rho_{k+1}, \ldots, \rho_{k'}$ are respectively the instantiations of $X_{k+1}, \ldots, X_{k'}$.

If in c we replace each argument by a valid instantiation, we get an *instantiated clause* noted $A_0(\boldsymbol{\rho_0}) \to A_1(\boldsymbol{\rho_1}) \ldots A_r(\boldsymbol{\rho_r})$ in which each $A_i(\boldsymbol{\rho_i})$ is an *instantiated predicate*.

A binary relation called *derive* and noted $\underset{G,w}{\Rightarrow}$ is defined on strings of instantiated predicates. If Γ_1 and Γ_2 are strings of instantiated predicates, we have

$$\Gamma_1 \; A_0(\boldsymbol{\rho_0}) \; \Gamma_2 \underset{G,w}{\Rightarrow} \Gamma_1 \; A_1(\boldsymbol{\rho_1}) \ldots A_m(\boldsymbol{\rho_m}) \; \Gamma_2$$

if and only if $A_0(\boldsymbol{\rho_0}) \to A_1(\boldsymbol{\rho_1}) \ldots A_m(\boldsymbol{\rho_m})$ is an instantiated clause.

The *(string) language* of a PRCG G is the set $\mathcal{L}(G) = \{w \mid S(\langle 0..|w|\rangle_w) \underset{G,w}{\overset{+}{\Rightarrow}} \varepsilon\}$.

In other words, an input string $w \in T^*$, $|w| = n$ is a *sentence* of G if and only if there exists a *complete derivation* which starts from $S(\langle 0..n\rangle)$ (the instantiation of the start predicate on the whole input text) and leads to the empty string (of instantiated predicates). The *parse forest* of w is the CFG whose axiom is $S(\langle 0..n\rangle)$ and whose productions are the instantiated clauses used in all complete derivations.[3]

We say that the arity of a PRCG is k, and we write k-PRCG, if and only if k is the maximum arity of its predicates ($k = \max_{A \in N} arity(A)$). We say that a k-PRCG is *simple*, we have a simple k-PRCG, if and only if each of its clause is

- *non-combinatorial*: the arguments of its RHS predicates are single variables;
- *non-erasing*: each variable which occur in its LHS (resp. RHS) also occurs in its RHS (resp. LHS);
- *linear*: there are no variables which occur more than once in its LHS and in its RHS.

The subclass of simple PRCGs is of importance since it is MCS and is the one equivalent to LCFRSs.

We say that a simple 2-PRCG clause of the form

$$\phi \to A_1(X_1, X_1') \ldots A_p(X_p, X_p') \; B_1(Y_1) \ldots B_q(Y_q)$$

in which the LHS is either of the form $B_0(\alpha)$ or $A_0(\alpha_0, \alpha_0')$ with $\alpha = \alpha_0 \alpha_0'$ is *well-balanced* if and only if α is a Dyck string w.r.t. the pairs (X_i, X_i') which play the role of parentheses, X_i is an open parenthese while X_i' is a closing parenthese. A simple 2-PRCG is *well-balanced* if its clauses are all well-balanced.

3   TAGs and MCTAGs vs. Simple PRCGs

In this Section we briefly present the algorithms of [10,5] with a slight modification due to our (simplified) vision[4] of TAGs and MCTAGs.

---

[3] Note that this parse forest has no terminal symbols (its language is the empty string).
[4] No substitution, no adjunction neither at the root nor at the foot node and mandatory adjunction on inside nodes.

3.1   Transforming TAGs into Simple 2-PRCGs

If we consider an auxiliary tree τ and the way it evolves until no more adjunction is possible, we realize that some properties of the final tree are already known on τ. The yield derived by the part of τ to the left (resp. to the right) of its spine are contiguous terminals and the *left yield* (produced by the left part) lies to the left of the *right yield* in any input string. Thus, for any elementary tree τ consider its m internal adjunction nodes. We decorate each such node η_i with two variables L_{η_i} and R_{η_i} $(1 \le i \le m)$ which are supposed to capture respectively the left and right yield of the trees that adjoined at η_i. Each terminal leaf has a single decoration which is its terminal label. Then, during a top-down left-to-right traversal of τ, we collect into a string σ_τ called *decoration string*, all such decorations. If τ is an auxiliary tree, let σ_τ^l and σ_τ^r be the part of σ_τ gathered before and after the traversal of the root of τ. To each elementary tree τ, we associate a simple 2-PRCG clause constructed as follows:

- its LHS is the predicate $S(\sigma_\tau)$ if τ is an initial S-tree or
- its LHS is the predicate $A(\sigma_\tau^l, \sigma_\tau^r)$ if τ is an auxiliary A-tree;
- its RHS is $\psi_1 \ldots \psi_i \ldots \psi_m$ with $\psi_i = A_i(L_{\eta_i}, R_{\eta_i})$ where A_i is the label of η_i.

Example 2. The following TAG

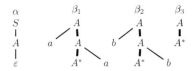

where α is its initial tree and β_1, β_2 and β_3 are its auxiliary trees, defines the language $\{ww \mid w \in \{a, b\}^*\}$ which is translated into the strongly equivalent simple 2-PRCG[5]

$$\begin{aligned} S(L_A R_A) &\to A(L_A, R_A) \\ A(aL_A, aR_A) &\to A(L_A, R_A) \\ A(bL_A, bR_A) &\to A(L_A, R_A) \\ A(\varepsilon, \varepsilon) &\to \varepsilon \end{aligned}$$

As an example, the arguments of the LHS predicate of the second clause have been gathered during the following traversal of β_1

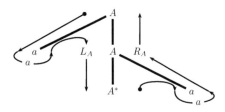

---

[5] Since there is no possible confusion, nodes are addressed directly by their labels.

It has been shown in [5] how the simple 2-PRCG generated by the previous algorithm can be turned into a normal form which contains at most two variables (the 2-var form) in each argument.[6] Since the standard parser of [11] has a worst case parse time complexity of $\mathcal{O}(|G|n^{k+v})$ for a k-RCG G in which v is the maximum number of variables per clause, we reach the classical $\mathcal{O}(n^6)$ parse time for TAGs.

3.2 Transforming Set-Local k-MCTAGs into Simple $2k$-PRCGs

The transformation of any set-local k-MCTAG into an equivalent simple $2k$-PRCG is based upon a generalization of the algorithm of Section 3.1. The difficulty is to identify the nodes where MCA operations can take place. This will be done in introducing the notion of *cover*.

Let us consider an elementary tree set γ. For each of its internal nodes η which is a (mandatory) adjunction node, we have to find a tree τ_i (in some auxiliary tree set β) which could be adjoined at η. A mapping noted ξ from each adjunction node of γ to an auxiliary tree τ_i (we have $\tau_i = \xi(\eta)$) in some auxiliary tree set β is called a *cover* (of γ). Note that different adjunction nodes of γ may be associated with trees that come from different tree sets. To be a cover, the total function ξ must fullfilled three constraints for each internal node η:

1. if $\xi(\eta) = \tau_i$, the label of η must be the label of the root of τ_i (only adjunction of A-trees are allowed on A-nodes);
2. if β is the tree set such that $\tau_i \in \beta$, then for each auxiliary tree $\tau_j \in \beta$ there exists an adjunction node of γ such that $\xi(\eta') = \tau_j$ (this constraint is the consequence of the definition of a MCA operation as the simultaneous adjunction of all its component trees in an auxiliary tree set) and
3. if η_1 and η_2 are two distinct nodes of γ s.t. $\xi(\eta_1) = \tau$ and $\xi(\eta_2) = \tau$ for some auxiliary tree τ in some auxiliary tree set β and if τ' is another tree of β then there exist two distinct nodes η'_1 and η'_2 of γ s.t. $\xi(\eta'_1) = \tau'$ and $\xi(\eta'_2) = \tau'$ (this constraint forces trees in an auxiliary tree set to be adjoined the same number of times during any MCA).

Note that empty auxiliary trees have exactly one cover, namely the empty cover.

Of course, for any given elementary tree set γ its cover is, in the general case, not unique (it may even not exist, and in this case γ can not be used in a derivation). However, for any elementary tree set in any given MCTAG, the number of its covers is bounded. Let $cover(\gamma)$ be the set of all the covers of the elementary tree set γ.

The generation of RCG clauses proceeds as follow. Each elementary tree set γ will yield $|cover(\gamma)|$ simple PRCG clauses, one for each cover ξ in $cover(\gamma)$. All these clauses will share the extraction of the decoration strings of γ. If γ is an initial tree set (a singleton), we associate a single decoration string σ_γ, while, if γ is an auxiliary tree set s.t. $|\gamma| = k$, we associate a sequence of $2k$

---

[6] Such a transformation, which is always possible, is based upon the fact that decoration strings are well-parenthesized (Dyck) string.

decoration strings $\sigma^l_{\gamma.1}, \sigma^r_{\gamma.1}, \ldots, \sigma^l_{\gamma.k}, \sigma^r_{\gamma.k}$, the pair $\sigma^l_{\gamma.h}, \sigma^r_{\gamma.h}$ in this sequence being respectively the left and right decoration strings of the (auxiliary) tree of rank h in γ. All $|cover(\gamma)|$ simple PRCG clauses will have the same LHS: $S(\sigma_\gamma)$ if γ is an initial tree set or $< \gamma > (\sigma^l_1, \sigma^r_1, \ldots, \sigma^l_k, \sigma^r_k)$ if γ is an auxiliary tree set. Now consider the RHS's of these clauses. Each ξ in $cover(\gamma)$ will produce a different RHS. For a given ξ, let $\beta_1, \ldots, \beta_m$ be the list of auxiliary tree sets that it selects. The RHS of the clause computed for ξ will be of the form $< \beta_1 > (L^1_1, R^1_1 \ldots, L^{|\beta_1|}_1, R^{|\beta_1|}_1) \ldots < \beta_m > (L^1_m, R^1_m \ldots, L^{|\beta_m|}_m, L^{|\beta_m|}_m)$ in which the L's and R's are the left and right variables that occur in the LHS decoration strings and that are associated to the adjunction nodes of γ. For each node η in γ let its two variables be L_η and R_η and let r be the rank of the auxiliary tree $\xi(\eta)$ in the auxiliary tree set β_j. We thus have $L_\eta = L^{\beta_j}_r$ and $R_\eta = R^{\beta_j}_r$.

Example 3. Let us consider the 3-MCTAG of Example 1. The decoration string σ_α of its initial tree set α is $\sigma_\alpha = L_{\alpha.1} R_{\alpha.1} L_{\alpha.2} R_{\alpha.2} L_{\alpha.2} R_{\alpha.2}$. It yields the three clauses

$$S(L_{\alpha.1} R_{\alpha.1} L_{\alpha.2} R_{\alpha.2} L_{\alpha.2} R_{\alpha.2}) \rightarrow < \beta_1 > (L_{\alpha.1}, R_{\alpha.1}, L_{\alpha.2}, R_{\alpha.2}, L_{\alpha.2}, R_{\alpha.2})$$
$$S(L_{\alpha.1} R_{\alpha.1} L_{\alpha.2} R_{\alpha.2} L_{\alpha.2} R_{\alpha.2}) \rightarrow < \beta_2 > (L_{\alpha.1}, R_{\alpha.1}, L_{\alpha.2}, R_{\alpha.2}, L_{\alpha.2}, R_{\alpha.2})$$
$$S(L_{\alpha.1} R_{\alpha.1} L_{\alpha.2} R_{\alpha.2} L_{\alpha.2} R_{\alpha.2}) \rightarrow < \beta_3 > ((L_{\alpha.1}, R_{\alpha.1}, L_{\alpha.2}, R_{\alpha.2}, L_{\alpha.2}, R_{\alpha.2})$$

since, for α there are three covers ξ_i, $1 \le i \le 3$

ξ_i	$\alpha.1$	$\alpha.2$	$\alpha.3$
	β_{i1}	β_{i2}	β_{i3}

The six decorations strings associated with the auxiliary tree sets β_i, $1 \le i \le 2$ have the form $\sigma^l_{\beta_{ik}} = aL_{\beta_{ik}.2}$ and $\sigma^r_{\beta_{ik}} = aR_{\beta_{ik}.2}$, for $1 \le k \le 3$. They yield the six clauses ($1 \le i \le 2$ and $1 \le j \le 3$)

$$< \beta_i > (aL_{\beta_{i1}.2}, R_{\beta_{i1}.2}, aL_{\beta_{i2}.2}, R_{\beta_{i2}.2}, aL_{\beta_{i3}.2}, R_{\beta_{i3}.2})$$
$$\rightarrow < \beta_j > (L_{\beta_{i1}.2}, R_{\beta_{i1}.2}, L_{\beta_{i2}.2}, R_{\beta_{i2}.2}, L_{\beta_{i3}.2}, R_{\beta_{i3}.2})$$

since for β_1 and β_2 there are the six covers ξ_{ij}, $1 \le i \le 2$ and $1 \le j \le 3$

ξ_{ij}	$\beta_{i1}.2$	$\beta_{i2}.2$	$\beta_{i3}.2$
	β_{j1}	β_{j2}	β_{j3}

Finally the six trivial decoration strings of β_3, i.e., $\sigma^l_{\beta_{3k}} = \sigma^r_{\beta_{3k}} = \varepsilon$ for $1 \le k \le 3$ yield the single clause $< \beta_3 > (\varepsilon, \varepsilon, \varepsilon, \varepsilon, \varepsilon, \varepsilon) \rightarrow \varepsilon$, since there are no covers for β_3.[7]

If we apply to this particular case the general formula of [11], we get for the generated simple 2k-PRCG G a worst case parse time complexity of $\mathcal{O}(|G|n^{2(k+v)})$ where v is the maximum number of adjunction nodes in an elementary tree set.

---

[7] We can note that this translation is not optimal in the sense that since β_1, β_2 and β_3 play a symetric role, the three predicates $< \beta_1 >$, $< \beta_2 >$ and $< \beta_3 >$ could be merged into a single predicate $< \beta_{123} >$.

However, in the TAG case, since the decoration strings are well parenthesized, the generated simple 2-PRCG can be transformed into an equivalent 2-var form in which the dependance w.r.t. to the number of variables (i.e., twice the number of adjunction nodes in an elementary tree) can always be reduced to four, leading to the famous $\mathcal{O}(n^6)$. Thus, we can wonder whether a similar transformation cannot be performed in the MCTAG case. Unfortunately, the answer is no. This comes from the fact that, in the general case, the decoration strings of elementary tree sets are so completely interlaced that it is not possible to isolate one MCA site, without isolating the others. The search of an optimal solution (in which the arity k may change) is still an open problem.

3.3   Simple 2-PRCGs vs. TAGs

We have seen in Section 3.1 how any TAG can be transformed into a strongly equivalent simple 2-PRCG. We will show below, thanks to two examples, that there exist simple 2-PRCGs which cannot be transformed into equivalent TAGs. This shows that the set of all simple 2-PRCLs strictly contains that of TALs.

Example 4. Consider the language $L = \{a_1^n b_1^m a_2^n b_2^m \mid n, m > 0\}$ which may be seen as a transducted copy of $a_1^n b_1^m$ into $a_2^n b_2^m$. It is not difficult to see that L is a TAL. But if we add the constraint that both the a's and b's must be related in such a way that they are well parenthesized ((a_1, a_2) and (b_1, b_2) are parenthesized), one can show that this additional constraint cannot be fulfilled by a TAG. However the following simple 2-PRCG in 2-var-form defines L. But we can note that the second clause is not well-balanced.

$$
\begin{aligned}
S(XY) &\rightarrow S_0(X, Y) \\
S_0(L_A L_B, R_A R_B) &\rightarrow A(L_A, R_A)\ B(L_B, R_B) \\
A(a_1 L_A, R_A a_2) &\rightarrow A(L_A, R_A) \\
A(\varepsilon, \varepsilon) &\rightarrow \varepsilon \\
B(b_1 L_B, R_B b_2) &\rightarrow B(L_B, R_B) \\
B(\varepsilon, \varepsilon) &\rightarrow \varepsilon
\end{aligned}
$$

Example 5. It can be shown thanks to the pumping lemma for LCFRLs [12] that the language $L_2 = \{a^m b^m c^n d^n e^m f^m g^n h^n \mid m, n > 0\}$ is not a TAL. However it is defined by the following simple 2-RCL in 2-var form. Once again, we can see that the second clause is not well-balanced.

$$
\begin{aligned}
S(XY) &\rightarrow [a^m b^m c^n d^n, e^m f^m g^n h^n](X, Y) \\
[a^m b^m c^n d^n, e^m f^m g^n h^n](XY, ZT) &\rightarrow [a^m b^m, e^m f^m](X, Z)\ [c^n d^n, g^n h^n](Y, T) \\
[a^m b^m, e^m f^m](aXb, eZf) &\rightarrow [a^m b^m, e^m f^m](X, Z) \\
[a^m b^m, e^m f^m](\varepsilon, \varepsilon) &\rightarrow \varepsilon \\
[c^n d^n, g^n h^n](cYd, gTh) &\rightarrow [c^n d^n, g^n h^n](Y, T) \\
[c^n d^n, g^n h^n](\varepsilon, \varepsilon) &\rightarrow \varepsilon
\end{aligned}
$$

In other words, simple 2-PRCGs are more powerful than TAGs, both from a weak and strong point of view.

If a simple 2-PRCG can be transformed into an equivalent simple 2-PRCG in 2-var form (not necessarily well-balanced), this PRCG can be parsed in $\mathcal{O}(n^6)$ time. However, there exist simple 2-PRCGs that cannot be transformed into equivalent 2-var form and thus cannot be parsed in $\mathcal{O}(n^6)$ (by the standard parsing algorithm).

Example 6. This is for example the case for a clause of the form

$$E(L_A L_B L_C L_D, R_C R_A R_D R_B) \to A(L_A, R_A)\ B(L_B, R_B)\ C(L_C, R_C)\ D(L_D, R_D)$$

for which there exist no non-empty substrings σ_1, σ_2, σ_3 and, σ_4, such that $\sigma_1\sigma_2 = L_A L_B L_C L_D$, $\sigma_3\sigma_4 = R_C R_D R_A R_B$ and σ_3 and σ_4 (or σ_4 and σ_3) contain all the closing patenthesizes of σ_1 and σ_2. Thus this clause, of parse time complexity $\mathcal{O}(n^{10})$, cannot be decomposed into more simple clauses of arity 2 with a number of variables less than 8.

4   Multi-Component TIGs

4.1   Definition

As mentioned in Section 2.1, a TIG is a restricted TAG where auxiliary trees must be either empty or left or right (auxiliary TIG tree for short) and adjunctions are disallowed on both root and foot nodes of auxiliary trees.

A Multi-Component TIG (MCTIG), is a MCTAG in which all trees in an auxiliary tree set are auxiliary TIG trees. Of course, if an adjunction node is on the spine of a left (resp. right) auxiliary tree, only a left (resp. right) auxiliary TIG tree or an empty tree can be adjoined at that node.

The maximum number k of auxiliary TIG trees in an auxiliary tree set is the *arity* of the MCTIG, we thus have a k-MCTIG.

4.2   TIGs are Equivalent to Simple 1-PRCGs

Of course, a 1-MCTIG is a TIG. We will first see how it is possible to transform a TIG into an equivalent simple 1-PRCG. In fact this transformation is a simplification of the algorithm shown in Section 3.1 wich transforms a TAG into an equivalent PRCG: in all cases an elementary TIG tree gives birth to a single decoration string during its top-down left-to-right traversal. The top-down traversal of the root node produces an empty symbol (since it is either an initial S-tree and there is no auxiliary S-trees or it is an auxiliary tree and adjunctions are not allowed at its root). The left-to-right traversal of leaf nodes labelled by a, $a \in T \cup \{\varepsilon\}$ produces the symbol a. The top-down (resp. bottom-up) traversal of an adjunction node η of label A produces the variable L_η (resp. R_η) if there exist left (resp. right) auxiliary A-trees or ε in the other cases.

Each non-empty elementary (initial or auxiliary) tree τ gives birth to a simple 1-PRCG clause c. Let σ be the decoration string of τ. If τ is an initial

S-tree, the LHS of c is $S(\sigma)$, while if τ is a left (resp. right) auxiliary A-tree, the LHS of c is $[A]^l(\sigma)$ (resp. $[A]^r(\sigma)$).

If τ is a non-empty elementary tree with $|\tau|$ adjunction nodes, the RHS of c has the form $\phi_{\eta_1}^l \phi_{\eta_1}^r \cdots \phi_{\eta_{|\tau|}}^l \phi_{\eta_{|\tau|}}^r$. Let us denote by A_i the label of the node η_i. If L_{η_i} (resp. R_{η_i}) is a variable in σ we have $\phi_{\eta_i}^l = [A_i]^l(L_{A_i})$ (resp. $\phi_{\eta_i}^r = [A_i]^r(R_{A_i})$) or if L_{η_i} (resp. R_{η_i}) is not (a variable) in σ we have $\phi_{\eta_i}^l = \varepsilon$ (resp. $\phi_{\eta_i}^r = \varepsilon$).

If τ is an empty auxiliary A-tree, it gives birth to the pair of clauses $[A]^l(\varepsilon) \to \varepsilon$ and $[A]^r(\varepsilon) \to \varepsilon$.[8]

Example 7. We can find below the elementary trees of a TIG[9] which defines the language $\{a^n b^n \mid n \geq 0\}$. To each tree is associated its decoration string and the associated generated clauses. We can note that the absence of right auxiliary tree in that TIG implies that there are no R-variables and thus no $[]^r$-predicates.

$$
\begin{array}{ccc}
\begin{array}{c} S \\ | \\ A \\ | \\ \varepsilon \end{array}
&
\begin{array}{c} A \\ {\diagup}\ {\diagdown} \\ a \quad A \\ {\diagup}\ {\diagdown} \\ b \quad A* \end{array}
&
\begin{array}{c} A \\ | \\ A* \end{array}
\\[2em]
\sigma = L_A & \sigma = a L_A b & \sigma = \varepsilon
\\[1em]
S(L_A) \to [A]^l(L_A) & [A]^l(a L_A b) \to [A]^l(L_A) & [A]^l(\varepsilon) \to \varepsilon
\end{array}
$$

In [13], Boullier has shown that any 1-RCG can always be transformed into an equivalent 1-RCG in 2-var form and can thus be parsed in $\mathcal{O}(n^3)$ time. This is an other way to show that TIGs can be parsed in $\mathcal{O}(n^3)$ time.

For space reasons, we will not show here how to perform the reverse conversion from a simple 1-PRCG to a TIG.[10] However, put together, these two conversion algorithms prove that simple 1-PRCGs are equivalent to TIGs and thus to 1-MCTIGs.

Before considering the general case of the k-MCTIGs we study, in this Section, the relationships between TAGs, 2-MCTIGs and Simple 2-PRCGs. We first show that 2-MCTIGs and Simple 2-PRCGs are two equivalent formalisms.

4.3   2-MCTIGs and Simple 2-PRCGs are Equivalent

To start, we show that a 2-MCTIG can be transformed into an equivalent simple 2-PRCG.

---

[8] Note that the first (resp. second) clause will never be used in any derivation if there is no left (resp. right) auxiliary A-trees. Of course, in that case, their generation may be skipped.

[9] Since there is no possible confusion, their nodes are addressed by their labels.

[10] The underlying idea is very simple: each clause is transformed into a (left) auxiliary tree: the root A_0 is the LHS predicate name; in A_0's (unique) argument, each sequence $u_i \in T*$ corresponds to leave nodes attached to the root, and each variable symbol $X_k \in V$ correspond to a subtree $\overset{X_k}{\underset{|}{}}$ attached to the root.

Let α be an initial tree set. By definition we have $|\alpha| = 1$ and τ is a S-tree if $\alpha = \{\tau\}$. The translation of α gives birth to a S-clause, where S is the unary start predicate. Let β be an auxiliary tree set. By definition we have $|\beta| \leq 2$. The translation of β gives birth to $[\beta]$-clauses, the predicate $[\beta]$ is unary if $|\beta| = 1$ or binay if $|\beta| = 2$.

We will only consider there the more intricate case of a binary auxiliary tree set, leaving the reader to figure out how the other (simpler) cases can be processed. Let $\beta = \{\tau_1, \tau_2\}$ be a binary auxiliary tree set with its two decoration strings σ_{τ_1} and σ_{τ_2}, and with ξ one of its cover in $cover(\beta)$ (see Section 3.2). Let η_1 and η_2 be a pair of adjunction nodes in β and let τ_1' and τ_2' be the two auxiliary TIG trees of some auxiliary tree set β' selected by ξ (we have $\tau_1' = \xi(\eta_1)$ and $\tau_2' = \xi(\eta_2)$).

If τ_1' is a left (resp. right) auxiliary tree, we erase the variable R_{η_1} (resp. L_{η_1}) from the (argument) string $(\sigma_{\tau_1}, \sigma_{\tau_2})$. Similarily, if τ_2' is a left (resp. right) auxiliary tree, we erase the variable R_{η_2} (resp. L_{η_2}) from the string $(\sigma_{\tau_1}, \sigma_{\tau_2})$. Let K_{η_1} and K_{η_2} the variables which have thus been kept.

In the RHS of our clause we generate a binary predicate $[\beta']$ whose two arguments are either (K_{η_1}, K_{η_2}) or (K_{η_2}, K_{η_1}). It is (K_{η_1}, K_{η_2}) if τ_1' is the first tree in the ordered tree set β' while it is (K_{η_2}, K_{η_1}) if τ_1' is the second tree in β'.

Of course, we perform the same operation for all the other nodes of τ_1 and τ_2. When all the values of ξ are exhausted, we have built a RHS say ϕ_ξ and a new argument string $(\sigma_{\tau_1}', \sigma_{\tau_2}')$. This yields the clause $[\beta](\sigma_{\tau_1}', \sigma_{\tau_2}') \to \phi_\xi$.

Example 8. We can find below the elementary trees of a 2-MCTIG which defines the language $\{a^n b^n c^n d^n \mid n > 0\}$. The various ξ's are represented by dashed arrows. Below each tree set we display its decoration string. Since this grammar is a left 2-MCTIG, there are no R-variables. Below each decoration string we show the generated clauses.

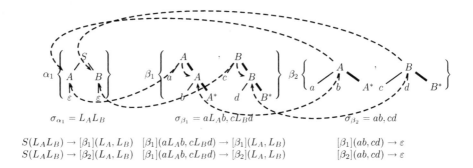

$$\sigma_{\alpha_1} = L_A L_B \qquad\qquad \sigma_{\beta_1} = aL_A b, cL_B d \qquad\qquad \sigma_{\beta_2} = ab, cd$$

$S(L_A L_B) \to [\beta_1](L_A, L_B)$ $\quad [\beta_1](aL_A b, cL_B d) \to [\beta_1](L_A, L_B)$ $\qquad [\beta_1](ab, cd) \to \varepsilon$

$S(L_A L_B) \to [\beta_2](L_A, L_B)$ $\quad [\beta_1](aL_A b, cL_B d) \to [\beta_2](L_A, L_B)$ $\qquad [\beta_2](ab, cd) \to \varepsilon$

We can note that this translation is not optimal since the two predicates $[\beta_1]$ and $[\beta_2]$ play the same role, the can be merged and thus led to a simple 2-PRCG which only contains three clauses.

For space reasons we will not show here how to perform the reverse conversion from a simple 2-PRCG to a 2-MCTIG.[11] However, put together, these two conversion algorithms prove that simple 2-PRCGs are equivalent to 2-MCTIGs.

4.4 TAGs vs. 2-MCTIGs

In this section we show that any TAG can be transformed into an equivalent 2-MCTIG, but that the converse is not true.

Of course, to show that a TAG can be transformed into an equivalent 2-MCTIG, we can perform a TAG to PRCG transformation followed by a PRCG to MCTIG transformation. But below we propose a direct transformation, that easily extends to k-MCTAGs and $2k$-MCTIGs (see Section 4.6).

Let σ be a TAG tree decoration string. This decoration string is either associated to an initial TAG tree or associated to the left part or the right part of an auxiliary TAG tree. As usual, two variables L_η and R_η are associated to an adjunction node η and will occur in σ if that node is not on the spine. If η is on the spine of an auxiliary tree β, only L_η will occur in the left decoration string of β while only R_η will occur in the right decoration string of β. Now, we rewrite σ in changing each occurrence of L_η (resp. R_η) by A_l (resp. A_r) if A is the label of η.

This process will transform each initial TAG tree (a S-tree) into an initial tree set of cardinality 1 (this single tree is a S-tree) of the 2-MCTIG.

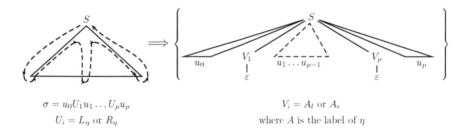

$$\sigma = u_0 U_1 u_1 \ldots U_p u_p$$
$$U_i = L_\eta \text{ or } R_\eta$$

$$V_i = A_l \text{ or } A_r$$
$$\text{where } A \text{ is the label of } \eta$$

It will also transform each auxiliary TAG tree (say an A-tree) into an auxiliary ordered tree set of cardinality 2 (an A_l-tree and an A_r-tree) of the 2-MCTIG. The A_l-tree (resp. A_r-tree) is built by the part of the auxiliary TAG tree to the left (resp. right) of its spine and by the spine itself.

---

[11] Informally, it is a generalization of the transformation sketched in footnote 10. For a given clause, we first rename all variables (i.e., all RHS arguments) such that the first argument of predicate A_i is A_i^1 and its second argument is A_i^2. Then, if the LHS predicate A_0 is unary, it gives birth to an A_0^1-rooted left auxiliary tree in the same way as in footnote 10; if A_0 is binary, it gives birth to an ordered set of two trees rooted by A_0^1 (resp. A_0^2), and built similarily, but using only A_i^1s (resp. A_i^2s). This supposes that the 2-PRCG has been first transformed into a strongly equivalent grammar in a so-called *LR-form* (A_i^1s are only in A_0's first argument and A_i^2s only in A_0's second argument), which can be proven always possible.

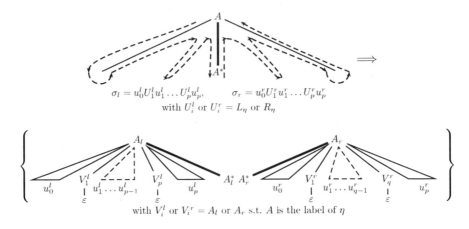

$$\sigma_l = u_0^l U_1^l u_1^l \ldots U_p^l u_p^l, \qquad \sigma_r = u_0^r U_1^r u_1^r \ldots U_p^r u_p^r$$
$$\text{with } U_i^l \text{ or } U_i^r = L_\eta \text{ or } R_\eta$$

with V_i^l or $V_i^r = A_l$ or A_r s.t. A is the label of η

Example 9. For example the TAG $\begin{smallmatrix}S\\|\\A\\|\\\varepsilon\end{smallmatrix}$, $a\!\!\diagup\!\!\begin{smallmatrix}A\\|\\A\\|\\b\end{smallmatrix}\!\!\diagdown\!\!\begin{smallmatrix}\\ \\A^*\end{smallmatrix} d$ and $\begin{smallmatrix}A\\|\\A^*\end{smallmatrix}$ which defines the language $a^n b^n c^n d^n$, $n > 0$ is transformed into the 2-MCTIG

$$\left\{\begin{smallmatrix}S\\\diagup\diagdown\\A_l\quad A_r\\|\qquad|\\\varepsilon\qquad\varepsilon\end{smallmatrix}\right\}\left\{\,a\diagup\!\!\begin{smallmatrix}A_l\\\diagup|\diagdown\\A_l\quad b\quad A_l^*\\|\\\varepsilon\end{smallmatrix}\quad A_r^*\diagdown\!\! c\diagup\!\!\begin{smallmatrix}A_r\\\diagup|\diagdown\\A_r\\|\\\varepsilon\end{smallmatrix}\!\!\diagdown d\right\}\left\{\begin{smallmatrix}A_l\quad A_r\\|\qquad|\\A_l^*\quad A_r^*\end{smallmatrix}\right\}$$

At this point, we know that any TAG can be transformed into a strongly equivalent simple 2-PRCG, and that simple 2-PRCGs and 2-MCTIGs are two equivalent formalisms. However, we have exhibited an example that shows that there are languages which can be defined by 2-MCTIG (or equivalently by simple 2-PRCGs) but which cannot be defined by TAGs.

4.5   2k-MCTIGs Are Equivalent to Simple 2k-PRCGs

We have already studied in Section 4.3 the case $k = 1$ which has been distinguished because of the particular importance of TAGs (i.e., 1-MCTAGs). In the general case, the results are very similar.

First of all, h-MCTIGs and simple h-PRCGs are two equivalent formalisms, and the arity h remains the same. On the one hand, to show that a h-MCTIG can be transformed into an equivalent simple h-PRCG is only a trivial generalization of what we have seen in Section 4.3. On the other hand, to show that a simple h-PRCG can be transformed into an equivalent h-MCTIG is also only a trivial generalization of the algorithm we have evoked in Section 4.3.

4.6   k-MCTAGs vs. 2k-MCTIGs

It is easy to show that a k-MCTAG can be transformed into an equivalent $2k$-MCTIG either indirectly by first using the transformation from MCTAG to simple PRCG of Section 3.2, followed by the transformation from a simple PRCG to a MCTIG mentioned in the same section. However, this transformation can be direct in generalizing the algorithm depicted in Section 4.4.

As for 2-MCTIG (and simple 2-PRCG), the converse is not true: $2k$-MCTIGs (and simple $2k$-PRCGs) are more powerful than k-MCTAGs.

However, MCTAGs, MCTIGs and simple PRCGs define the same class of languages since, by definition, a h-MCTIG may be seen as a h-MCTAG.

5 Conclusion: Hierarchies Comparison

Since k-MCTIGs and simple k-PRCGs are equivalent formalism, the hierarchy w.r.t. k of the class of languages defined by k-PRCGs, implies an identical hierarchy on the class of languages defined by k-MCTIGs. In particular let L_i be the counting language for i ($L_i = \{a_1^n \ldots a_i^n \mid n \geq 0\}$), the languages L_{2h-1} and L_{2h} can be defined by a h-MCTIG while $L_{2h+1}, L_{2h+2}, \ldots$ cannot. If we consider the class of MCTAGs, this figure slightly differs since the languages $L_{4h-3}, L_{4h-2}, L_{4h-1}$ and L_{4h} can be defined by a h-MCTAG while $L_{4h+1}, L_{4h+2}, \ldots$ cannot. This example shows that MCTAGs are more coarse-grained than MCTIGs in the sense that a single increment in the value of k delivers more (formal) power than the same increment in MCTIGs. In other words, for these three formalisms, in a $k+1$-class there exist languages that cannot be define by the corresponding k-class.

The following diagram summarizes, for each value of k and for each grammar class, in which other grammar class it can be transformed.

k	k-MCTAG	k-MCTIG	Simple k-PRCG
	(TAG)	(TIG)	(CFG)

The direction of dashed arrows shows a language inclusion.

This diagram shows that for each value of k, k-MCTIGs and simple k-PRCGs are equivalent formalisms. But, with MCTAGs, the relation is not so simple. We know that a h-MCTAG can be transformed into an equivalent simple $2h$-PRCG (and hence into an equivalent $2h$-MCTIG), but we know that $2h$-PRCGs (or $2h$-MCTIG) are more powerful than h-MCTAGs. On the other hand we know that a $2h$-MCTIG is, by definition a $2h$-MCTAG.

We can wonder whether there exists a value l with $h < l \leq 2h$ such that the two formalisms l-MCTAGs and a $2h$-MCTIGs are equivalent? The answer is no. Assume that such a value l exists. In that case we know that the counting language L_{4l} can be defined by a l-MCTAG but L_{4l} can only be defined by a k-MCTIG in which $k \geq 2l$. This is in contradiction with the fact that $h < l$.

From an operational point of view, the equivalence results presented in this paper show that we can parse k-MCTIGs with a worse-case parsing complexity of $O(|G|n^{k+v})$, where G is the equivalent PRCG and v the maximum number of adjunction nodes in an elementary tree set.

References

1. Joshi, A.K.: In: Dowty, D., Karttunen, L., Zwicky, A. (eds.) How much context-sensitivity is necessary for characterizing structural descriptions — Tree Adjoining Grammars. Cambridge University Press, New-York (1985)
2. Weir, D.: Characterizing Mildly Context-Sensitive Grammar Formalisms. PhD thesis, University of Pennsylvania, Philadelphia, PA (1988)
3. Joshi, A.K.: An introduction to tree adjoining grammars. In: Manaster-Ramer, A. (ed.) Mathematics of Language, pp. 87–114. John Benjamins, Amsterdam (1987)
4. Vijay-Shanker, K., Weir, D., Joshi, A.K.: Characterizing structural descriptions produced by various grammatical formalisms. In: Proceedings of the 25th Meeting of the Association for Computational Linguistics (ACL 1987), pp. 104–111. Stanford University, CA (1987)
5. Boullier, P.: On TAG parsing. Traitement Automatique des Langues (T.A.L.) 41(3), 759–793 (2000)
6. Boullier, P.: On Multicomponent TAG parsing. In: 6me conference annuelle sur le Traitement Automatique des Langues Naturelles (TALN 1999), Cargse, Corse, France, pp. 321–326 (July 1999); see also Research Report No. 3668 at http://www.inria.fr/RRRT/RR-3668.html, INRIA-Rocquencourt, France, 39 pages (April 1999)
7. Barthélemy, F., Boullier, P., Deschamp, P., de la Clergerie, É.: Guided parsing of range concatenation languages. In: Proceedings of the 39th Annual Meeting of the Association for Computational Linguistics (ACL 2001), pp. 42–49. University of Toulouse, France (July 2001)
8. Schabes, Y., Waters, R.C.: Tree insertion grammar: A cubic-time, parsable formalism that lexicalizes context-free grammar without changing the trees produced. Computational Linguistics 21 (1994)
9. Joshi, A.K., Levy, L., Takahashi, M.: Tree adjunct grammars. Journal of Computer and System Sciences 10, 136–163 (1975)
10. Boullier, P.: A generalization of mildly context-sensitive formalisms. In: Proceedings of TAG+4, pp. 17–20. University of Pennsylvania, Philadelphia (August 1998)
11. Boullier, P.: Range Concatenation Grammars. In: Bunt, H., Carroll, J., Satta, G. (eds.) New Developments in Parsing Technology. Text, Speech and Language Technology, vol. 23, pp. 269–289. Kluwer Academic Publishers, Dordrecht (2004)
12. Seki, H., Matsumura, T., Fujii, M., Kasami, T.: On multiple context-free grammars. Theoretical Computer Science 88, 191–229 (1991)
13. Boullier, P.: A cubic time extension of context-free grammars. Grammars 3(2/3), 111–131 (2000)

A Grammar Correction Algorithm
Deep Parsing and Minimal Corrections
for a Grammar Checker

Lionel Clément[1], Kim Gerdes[2], and Renaud Marlet[3]

[1] LaBRI, Université Bordeaux 1
[2] ILPGA, LPP, Sorbonne Nouvelle
[3] LaBRI, INRIA Bordeaux – Sud-Ouest

Abstract. This article presents the central algorithm of an open system for grammar checking, based on deep parsing. The grammatical specification is a context-free grammar with flat feature structures. After a shared-forest analysis where feature agreement constraints are relaxed, error detection globally minimizes the number of corrections and alternative correct sentences are automatically proposed.

1 Introduction

Grammar checkers are among the most common NLP technologies used by the general public. Yet they have attracted comparatively little research, at least w.r.t. the number of publications. There are most likely various reasons for this. First, from a practical point of view, the systems appear to be highly dependent on the language they correct. In addition, a large part of grammatical definitions is a collection of idiosyncratic errors that may depend on the language of the user and his level of knowledge. This task is difficult to automate due to the unavailability of large error corpora. Finally, the usefulness of such a system relies heavily on its integration into a word processor, and a grammar checker can easily be made available to the public only since the rise of OpenOffice. It is now conceivable that a research community forms around grammar checking, openly sharing resources and results, just as in other areas of NLP. There may also be deeper reasons: linguistics has taken a long time to define itself as a science of the language that is actually spoken, and normativity is no longer at the center of interest, although some research on deviations from the norm (in sociolinguistics, psycholinguistics, lexicology, or in research on foreign language teaching) can have an indirect interest for the development of error grammars.

After showing the necessity for quality grammar checking based on deep syntactic analysis, as opposed to the frequently used superficial and local grammars, we present the algorithm at the heart of our system. We show how to detect, locate, and order minimal correction proposals, using a tweaked-up context free grammar parser. This is the central module of the open grammar checker that we are developing, using, for the moment, a French grammar that is also used here for illustration purposes.

P. de Groote, M. Egg, and L. Kallmeyer (Eds.): FG 2009, LNAI 5591, pp. 47–63, 2011.

2   How to Check Grammars

A grammar checker has two main functions:

- It notifies the user of possibly incorrect sentences (or fragments).
- It proposes corrections, possibly with a "linguistic" explanation of the error.

In practice, additional properties also are desirable:

- It should be fast: checking should be perform as the user types.
- It should minimise noise: too many false alarms would make it unusable.
- It should minimise silence: only few errors should remain unnoticed.

2.1   Minimal Correction Cost

Uszkoreit, quoted in [1], distinguishes 4 steps in the grammar correction process: (1) identification of possibly ungrammatical segments, (2) identification of the possibly infringed constraints, (3) identification of the possible source of the error, and (4) construction and ordering of the correct alternatives.

In this paper, we focus on this last point. The problem is to find alternative correct sentences in the *neighborhood* of an incorrect sentence and to order them by *plausibility*. We show how to model the plausibility of a correction in terms of its *minimal correction cost*.

The notion of minimality of a proposed correction is non trivial and different definitions can be given. Consider the following example, keeping in mind that French has an agreement in gender and number between the subject and the predicative adjective:

Example 1

(a) * Les cheval blanc sont arrogants.
   The-⟨*plur,masc/fem*⟩ horse-⟨*sing,masc*⟩ white-⟨*sing,masc*⟩ are-⟨*plur,3rd person*⟩ arrogant-⟨*plur,masc*⟩
(b) Le cheval blanc est arrogant.
   The-⟨*sing,masc*⟩ horse-⟨*sing,masc*⟩ white-⟨*sing,masc*⟩ is-⟨*sing,3rd person*⟩ arrogant-⟨*sing,masc*⟩
(c) Les chevaux blancs sont arrogants.
   The-⟨*plur,masc/fem*⟩ horses-⟨*plur,masc*⟩ white-⟨*plur,masc*⟩ are-⟨*plur,3rd person*⟩ arrogant-⟨*plur,masc*⟩

With a local decision principle, the subject noun phrase "les cheval blanc" has a best correction with a singular determiner ("le" instead of "les"). For the whole sentence, we end up with 3 singular words and 2 plural words, which tips the balance towards correction (b). With a global decision principle, however, we started out with 5 words that needed number agreement: 3 are plural, 2 are singular. The optimal correction, minimal in the number of words to change, is therefore (c). This shows that the correction is best taken globally, as local decisions can imply iterative suboptimal proposals.

Two proximity models allow for plausible corrections: we can count the number of words to be changed, or the number of features to be changed. Although the two approaches often agree, they are incomparable:

Example 2

(a) * C'est une histoire de cliente arrivée mécontent mais repartis satisfaits.
It is a story of client-$\langle sing,fem \rangle$ arrived-$\langle sing,fem \rangle$ unhappy-$\langle sing,masc \rangle$ but left-$\langle plur,masc \rangle$ satisfied-$\langle plur,masc \rangle$

(b) C'est une histoire de client arrivé mécontent mais reparti satisfait.
It is a story of client-$\langle sing,masc \rangle$ arrived-$\langle sing,masc \rangle$ unhappy-$\langle sing,masc \rangle$ but left-$\langle plur,masc \rangle$ satisfied-$\langle plur,masc \rangle$

(c) C'est une histoire de cliente arrivée mécontente mais repartie satisfaite.
It is a story of client-$\langle sing,fem \rangle$ arrived-$\langle sing,fem \rangle$ unhappy-$\langle sing,fem \rangle$ but left-$\langle sing,fem \rangle$ satisfied-$\langle sing,fem \rangle$

(d) C'est une histoire de clients arrivés mécontents mais repartis satisfaits.
It is a story of client-$\langle plur,masc \rangle$ arrived-$\langle plur,masc \rangle$ unhappy-$\langle plur,masc \rangle$ but left-$\langle plur,masc \rangle$ satisfied-$\langle plur,masc \rangle$
"It's a story of client(s) that came in unhappy but that left satisfied."

In (a), the "client" noun phrase is rather masculine, 3 votes against 2, and rather singular, equally 3 votes against 2. Minimizing the number of features to correct affects 4 occurrences of features and 4 words (b). But two alternatives are more economical w.r.t. the number of corrected words: (c) and (d) both correct 5 features but only 3 words. Our algorithm chooses the feature minimization, which seems more plausible from a "cognitive" point of view. Moreover, the word minimization choice can be encoded in our algorithm as well.

2.2   Positive Grammar vs. Negative Grammar

We can identify incorrect sentences with a *positive* grammar, generating all grammatical sentences, or with a *negative* grammar, describing ungrammaticalities.

However, we believe that a complete grammar checker actually needs both a positive and a negative grammar, for different types of error detection. The global structure of the sentence and the agreement phenomena is easier to described positively, because there are more ungrammatical variations than grammatical structures and agreement possibilities. Yet, other types of errors do not integrate as easily into a positive grammar. Common stylistic errors like barbarisms (e.g., the Gallicism 'to assist to a meeting') or pleonasms (e.g., 'free gift') are much more easily detected negatively by specific rules.

This separation in two types of grammars can also be motivated from linguistic or engineering standpoints: the positive grammar describes the basic general language properties that are independent of the speaker, whereas the negative grammar can be adapted for the user of the system, e.g., to detect false cognates that are specific to a language pair (e.g., 'affair' vs "affaire").

Although we currently only support a positive grammar, we plan to support a negative grammar as well, which would be based on deep syntactic information gathered during the (relaxed) positive grammar parsing.

2.3   Shallow and Deep Parsing

Grammar checking can be done with a variety of tools: part-of-speech tagging, chunking, regular expressions, unification grammars, and so on. The more an analysis is superficial, the more it is prone to noise and silence — not only because syntactic information is partial, but also because these tools have a high error rate on whole sentences. Shallow parsing is thus confined to negative grammars, where rule locality reduces these effects [2].

Most grammar checkers (e.g., for French: Antidote, Cordial, Microsoft Office, and Prolexis) are commercial products whose functioning is opaque. Although the parsing depth can sometime be known or guessed, the actual computation remains undisclosed. LanguageTool [3] is one of the few open grammar checkers, integrated in the upcoming version of OpenOffice (3.1). The grammar correction infrastructure is based on a part-of-speech tagger and on local rules built with regular expressions. More than 1600 rules are available for French. Consider:

Example 3

a * Le chien de mes voisins mordent.
   The dog of my neighbors bite.
b Les voitures qu'il a font du bruit.
   The cars that he has make noise.

With its local rules, LanguageTool remains silent on (a) and detects two false errors in (b). Microsoft Office [4], which seemingly makes a deeper analysis, flags the error in (a), but asks, wrongly, to replace "font" ('make') by "fait" ('made') in (b). In order to verify the global structure and the (possibly long-distance) agreements, we cannot do without a deep and complete analysis of the sentence.

In the rest of this article, we describe the deep parsing as well as the error detection and correction which form the central part of our grammar checker.

3   Positive Error Parsing and Correction

In short, the correction process of our checker is as follows. A sentence is first segmented into a lattice of inflected forms, i.e., a directed acyclic graph (DAG) with unique initial and final nodes, to which we add extra lemmas representing plausible substitutions (e.g., homophones). Figure 1a shows a simple DAG produced by the the lexical analysis. In French, "pomme de terre" and "terre cuite" are idioms ('potato' and 'terracotta') as well as phrases ('apple of earth' and 'cooked earth'). Besides, the lexer also gives two analyses to words like "du": as a partitive determiner and as a contraction of the preposition "de" ('of') and the definite determiner "le" ('the').

The parser then constructs a shared forest of analyses ignoring agreement phenomena. Next, a bottom-up pass through the forest attributes to alternative parses the cost of the minimal correction that satisfy the agreements, by modifying features, or of using substituted lemmas. A sentence that has a minimal cost of zero is grammatical; otherwise, it is ungrammatical and a top-down pass determines the inflections and substitutions that reconstruct correct sentences.

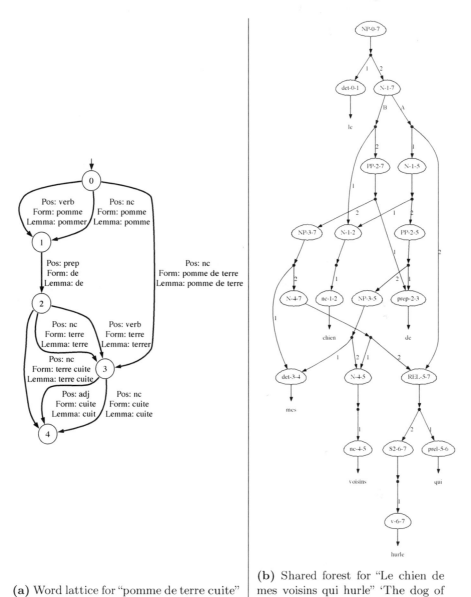

(a) Word lattice for "pomme de terre cuite"

(b) Shared forest for "Le chien de mes voisins qui hurle" 'The dog of my neighbours that howls'

Fig. 1. Examples of lexical and syntactic analyses

3.1 Lexical Correction

The notion of proximity significantly applies to word level: the lexicon has to define what forms can correct others and at what price.

A first type of corrections is based on inflection, i.e., different word forms of the same lemma, e.g., "jolie" ('pretty'-⟨*fem*⟩) for "joli" ('pretty'-⟨*masc*⟩). But not all features may vary freely, e.g., we can get the plural form of "crayon" ('pencil'-$\langle sing, masc \rangle$) but we cannot make it feminine; it somehow has an infinite cost.

A second type of corrections corresponds to plausible substitutions of lemmas, in particular homophones: "on" ('one') / "ont" ('have'), "est" ('is') / "ait" ('has') / "ai" ('have'), "à" ('to') / "a" ('has'), etc. As categories vary in this case, contrary to the previous situation with inflection, the parsing of these corrections adds additional alternative analyses, adding to the complexity of the overall process. The addition of such alternative lemmas therefore has to be done carefully.

Such corrections might first seem as an intrusion on the field of negative grammars, but the integration of similar lemmas is coherent with our goal: we want to include in the neighborhood of correct sentences (positive grammar) the most plausible incorrect corresponding sentences. Besides, the choice of the grouping of words under the same lemma is to a certain extent dependent on linguistic and theoretical considerations. For example, the frequently confused relative pronouns "que" ('which'-$\langle accusative \rangle$) and "dont" ('of which'-$\langle genitive \rangle$) can be seen either as different lemmas, or as a common *pronoun lemma* with different values for the case feature.

At the border between positive and negative grammar, we use a similar procedure for prepositions that are complements of verbs: we encode frequently subcategorized prepositions ("à", "de", "par", ...) as variations governed by a feature. Alternatively, we could have placed the detection of theses cases in the negative grammar. The choice between the two grammars depends on the difficulty to describe the possible error. Also, an advantage of expressing a phenomenon in the positive grammar is to automatically provide correction proposals.

For the positive grammar that we discuss here, the correction cost can depend on the type of lemma, on the type of forms, and may even be user-specific. It constitutes an independent resource. We thus have three lexical resources[1]:

- a classical *inflectional lexicon*, containing the lemmas, their word forms and their features constraints,
- a *substitution lexicon*, listing substitution groups of different lemmas/forms,
- a *model of proximity*, associating substitution costs to features and lemmas.

Our inflectional lexicon has the following extensional format:

```
# Form     Lemma      Categories and features
heureux    heureux    adj[gen=masc; nb=sing,plur]
portions   portion    nc[gen=fem!; nb=plur]
portions   porter     v[pers=1; nb=plur; mode=ind,subj; tps=imp]
```

In the case of different lemmas sharing the same form, we have an entry for each lemma. For each feature, a status concerning its possible values has to be provided: a value is either *prescribed* (i.e., compliant with the inflected form, thus assignable at a null cost), *plausible* (not covered by the inflected form but assignable to reach an agreement, at a non-null cost), or *forbidden* (not assignable, or equivalently, assignable at an infinite cost). These assignments

---

[1] Not counting a separate definition of features, shared by the lexicon and grammar.

constraints are specified by a list of values. By default all mentioned values are prescribed. The mark "!" forbids values other than those that are listed[2]. This syntax is light weight and furthermore allows for an easy static recognition of incoherence in the lexicon w.r.t. the feature definitions.

To a large extent, these kinds of lexicons can be build automatically [5], which is valuable for languages that are poor in free resources. In fact, our French lexicon is compiled from the Lefff, the free lexicon of French inflected forms, which was built on this basis. Our substitution lexicon, containing substitutable lemmas as "on|ont" or "a|à", stems mainly from the homophones of the *Lexique 3* [6]. The lexer systematically introduces these forms as alternatives into the lexical lattice before parsing.

Work in progress concerns a parameterizing tool for the cost model, including feature-dependant and substitution costs, as well as an extension based on a kind of Levenshtein distance to possibly balance correction costs according to lexicographic proximity. Presently, the cost model is a simple ad hoc program attributing costs following a fixed scheme: prescribed values have a cost of zero; plausible values have a cost of 1; and forbidden values have an infinite cost.

Substitutions, too, currently have a cost of 1. Contrary to other feature value assignments, this cost has to be paid as soon as the corresponding alternative is chosen. This can be modeled by an additional implicit feature, here named "$", with only one possible value: it thus always has to be assigned. N.B. All these arbitrary values are to be adjusted by a practical evaluation of the grammar.

Currently, most systems separate spelling correction and grammar checking. The user's best strategy is to correct spelling first. Our correction scheme allows a tighter integration of a spell checker and a grammar checker: correction proposals for unknown words can be added into in the lattice built by the lexer. A single operation can then correct wrong inflections like "chevals" (an imaginable but wrong plural form of "cheval" ('horse'), the correct plural form being "chevaux") or "faisez" (an imaginable but wrong second person plural form of "faire" ('to do'), the correct form being "faites").

3.2 The Principle of Constraint Relaxation

The notion of *proximity* for the correction of an ungrammatical sentence is subjective. The hypothesis that we make is that the proximity can be modeled with a positive grammar in which we relax some constraints. We thus obtain more analyses of the input sentence, and a cost assignment for the unsatisfied constraints measures the proximity to a correct sentence.

For most constraint-based formalisms, experiments have been done on constraint relaxations. Some formalisms as Property Grammars [7] even consider it as a key point of their approach. This idea also has been applied to HPSG [8]. And error ranking can be refined using weights representing feature importance [9]. For parsers of simpler grammars, constraint relaxation also has proved robust.

---

[2] Marks "?" and "#" (not exemplified here) identify respectively plausible and forbidden values, allowing more explicit or finer specifications.

Real-time grammar checking requires simple and efficient parsing mechanisms. Our system for parsing with constraint relaxation is based on a well known formalism: we use a feature-based context-free grammar, the features being flat and with finite values. Here is a simplified example of a rule:

```
sn[nb=N;gen=G;pers=3] -> det[nb=N;gen=G] sadj[nb=N;gen=G;
    type=anté]* nc[nb=N;gen=G] sadj[nb=N;gen=G;type=post]*
    pp[]? rel[nb=N; gen=G]? ;
```

We relax the agreement constraints related to features, but impose that the sentence be parsable by the bare (feature-free) context-free grammar; the correction cost computations are then performed on the context-free backbone. Note that the grammar can also be specifically written to relax some requirements and accept incorrect phrases with simple cases of structural incoherence, e.g., considering as grammar axioms not only sentences but also a few given phrases, or making some terms optional. Moreover, as explained earlier, we systematically add paths into the lexical lattice for words that are close to the actual input words (homophones or words that are often mistaken for one another).

We keep all structural ambiguities of our context-free analysis in a shared forest [10], using a variant of an Earley parser [11] that can directly handle Kleene stars and optional terms. The complexity is the same as for standard Earley parsing: the worst case complexity is cubic in the length of the input sentence. An example of such a shared parsing forest is given in figure 1b.

Moreover, our parser considers each word of a sentence as a potential starting point of the grammar axiom. We thus can construct partial analyses even for sentences and phrases that do not have a global analysis in our grammar, and flag local feature agreement errors in noun phrases, verb clusters, etc. This not only is useful for robustness, but also to signal meaningful errors even with a grammar with limited coverage.

We end up with two types of grammatical errors: *structural errors* (for which we do not find a single constituent structure) and *non structural errors* (for which unification fails) [12]. Note that adding an alternative to the substitution lexicon can transform a structural error into a non structural error. We can even manage to handle certain word order constraints by means of additional features and lemmas, e.g., for the anteposition and postposition of adjectives.

4 Signaling Errors

The text to analyze is assumed to be segmented into individual sentences[3]. As explained above, each such sentence is itself segmented by a lexer that constructs a lattice of inflected word-forms. Alternatives in this lattice are due to lexical ambiguities (including compounds) as well as possible word substitutions.

Parsing this lattice then constructs a shared forest of parse trees (see figure 1b). Features are entirely ignored at this stage; we only build the derivation

---

[3] Splitting heuristics based on punctuation provide a reasonable segmentation. We are studying a more robust technique that would exploit the fact that our parser is able to work on a stream of characters.

trees of the context-free skeleton of the grammar, with sharing. The resulting parse forest can be represented by a finite number of equations that belong to one of the following three types:

- $f = \{t_1, \ldots, t_m\}$ is a forest of m alternative trees, each tree being associated to the same non-terminal in the grammar.
- $t = \rho(f_1, \ldots, f_n)$ is the node of a tree, constructed using grammatical rule ρ, where n is the number of terms used in the right-hand side of ρ. (The number n refers to a rule instance, taking into account variants due to the option sign ? and the Kleene star *.)
- $t = l$ is a leaf node, constructed using lexical entry l.

The parser puts up with infinite analyses, as may occur (unwittingly) with rules such as "np -> np pp?". In this case, it generates cycles in the forest that can easily be removed. For what follows, the forest is assumed to be a lattice.

4.1   Error Mining

Looking for errors requires traversing the parse forest to determine parse alternatives with minimum correction costs. A forest that contains a rooted tree with a null correction cost is considered as error free. Otherwise, plausible corrections associated to the parse alternatives are ranked according to their cost.

A correction is represented as a *feature assignment* $\alpha = (u_\varphi)_{\varphi \in \Phi}$ such that:

- Φ is a set of features that each can be assigned a value.
- For each feature $\varphi \in \Phi$, $u_\varphi \in Dom(\varphi)$ is a feature value.

A feature assignment example is $\alpha = (\text{gender} \mapsto \text{masc}, \text{number} \mapsto \text{pl})$. Moreover, a *feature assignment cost* $\kappa = (c_\varphi)_{\varphi \in \Phi}$ is a family of *cost functions* such that:

- $c_\varphi : Dom(\varphi) \to \mathbb{N} \cup \{\infty\}$ associates an cost $c_\varphi(u_\varphi)$ to a feature value u_φ.

An example of a feature assignment cost for a masculine singular adjective is $\kappa = (\text{gender} \mapsto (\text{masc} \mapsto 0, \text{fem} \mapsto 1), \text{number} \mapsto (\text{sg} \mapsto 0, \text{pl} \mapsto 1))$. An infinite cost corresponds to an impossible feature assignment, such as imposing the masculine gender to 'mother' or, in French, imposing a feminine gender to "crayon" ('pen').

Last, to perform the feature assignment $\alpha = (u_\varphi)_{\varphi \in \Phi}$ with the feature assignment cost $\kappa = (c_\varphi)_{\varphi \in \Phi}$ has a total cost of $\kappa(\alpha) = \sum_{\varphi \in \Phi} c_\varphi(u_\varphi)$. For instance, with α and κ as defined above, the total cost is $\kappa(\alpha) = 0 + 1 = 1$.

To look for corrections in the forest and rank them, we investigate the feature assignment costs of parse alternatives. There can be an exponential number of such alternatives (w.r.t. the sentence length). In the rest of this subsection, we consider that we scan all these alternatives; in the following subsection, we show how to reduce and control this potential combinatorial explosion. Still, even though we examine each alternative, we exploit the sharing in the forest to factorize the computation of alternative feature assignment costs.

For each node t or forest f in the parse structure, we compute a set of *feature assignment costs* $(\kappa_i)_{i \in I} = ((c_{i,\varphi})_{\varphi \in \Phi_i})_{i \in I}$ representing the set of minimum costs for using t or f in an alternative parse i, for the different feature assignments. This computation operates bottom-up (from leafs to the roots) as follows[4].

(a) Let $t = l = s[\varphi = ...]_{\varphi \in \Phi}$ be a leaf in the parse forest. It represents a single alternative. Its feature assignment cost $(c_\varphi)_{\varphi \in \Phi}$ is defined by the cost model (cf. §3.1).

(b) Let $f = (t_i)_{i \in I}$ be a forest. Each alternative i is itself based on parse alternatives $j \in J_i$, whose feature assignment costs are $((c_{i,j,\varphi})_{\varphi \in \Phi_{i,j}})_{j \in J_i}$. The alternative feature assignment costs for f are thus $((c_{i,j,\varphi})_{\varphi \in \Phi_{i,j}})_{i \in I, j \in J_i}$, i.e., the set union of all alternative costs. (No global decision regarding assignment costs can be taken at this stage.)

(c) Let $t = \rho(f_i)_{i \in I}$ be a node built on the grammar rule ρ, $((c_{i,j,\varphi})_{\varphi \in \Phi_{i,j}})_{j \in J_i}$ be the alternative feature assignment costs of forests f_i, $c_0[\varphi = x_{0,\varphi}]_{\varphi \in \Phi_0}$ be the left-hand side of the instance of ρ, and $c_i[\varphi = x_{i,\varphi}]_{\varphi \in \Phi_i}$ be the term corresponding to f_i in the right-hand side of ρ.

1. For each forest f_i and each alternative cost $(c_{i,j,\varphi})_{\varphi \in \Phi_{i,j}}$ in f_i, we rule on features $\varphi \in \Phi_{i,j}$ that are not propagated in ρ for lack of equations of the form "$\varphi = x$" in $c_i[...]$. The minimum assignment cost of such a feature φ, for this alternative (i, j), is $\min(c_{i,j,\varphi}) = \min_{u \in Dom(\varphi)} c_{i,j,\varphi}(u)$. E.g., with cost κ as above, we have $\min(c_{number}) = \min(0, 1) = 0$.

   As it is not possible to rule on propagated features, as going up the forest may add new constraints, it is not yet possible to tell which alternative costs less. The minimum assignment cost $\min(c_{i,j,\varphi})$ must thus be preserved for when the other features of alternative (i, j) are evaluated.

   For this, we use the special feature named $ (that is also used in the substitution lexicon, cf. §3.1), that we systematically propagate up the forest: the $ feature is *implicitly* considered as present at each non-terminal occurrence, associated with the same feature variable, named $x_\$$. In other words, each rule $s_0[...] \to s_1[...] \ldots s_n[...]$ actually represents the rule $s_0[...; \$ = x_\$] \to s_1[...; \$ = x_\$] \ldots s_n[...; \$ = x_\$]$. For the rest of the computations, ruling on non-propagated features amounts to replacing each feature assignment cost $(c_{i,j,\varphi})_{\varphi \in \Phi_{i,j}}$ by $(c'_{i,j,\varphi})_{\varphi \in \Phi'_{i,j}}$ where:
   - The set of propagated features of alternative (i, j) is: $\Phi'_{i,j} = \Phi_{i,j} \cap \Phi_i$.
   - The set of non-propagated features of alternative (i, j) is: $\Phi_{i,j} \setminus \Phi_i$.
   - For propagated features $\varphi \in \Phi'_{i,j}$, except $, we have $c'_{i,j,\varphi} = c_{i,j,\varphi}$.
   - Feature $ accumulates the minimum costs of non-propagated features of alternative (i, j): $c'_{i,j,\$} = c_{i,j,\$} + \sum_{\varphi \in \Phi_{i,j} \setminus \Phi_i} \min(c_{i,j,\varphi})$.

---

[4] For lack of space, we do not describe here the treatment of a number of grammar constructs: filtering (equivalent to $=_c$ in LFG), constant equality constraint ("$\varphi = v$" in the left- or right-hand side of a rule), joint constant and variable equality constraint (written "$\varphi = x \& v$"). The "$\varphi = x$" case, that we describe here, is at the heart of the minimal correction algorithm.

2. We then construct the *agreement cost* of the variables of rule ρ. We consider each alternative $\bar{\jmath} = (j_i)_{i \in I} \in \bar{J} = \prod_{i \in I} J_i$, i.e., all combinations of an alternative from each forest f_i. (See §4.2 for combinatorial reduction.) For each feature variable x, whose value domain is $Dom(x)$, in the set X_r of feature variables occurring on the right-hand side of ρ, we define a cost function $c_x : Dom(x) \to \mathbb{N} \cup \{\infty\}$ that represents the sum of the feature assignment costs that are associated to x in the different alternatives j_i in which x occurs in $s_i[\ldots]$. In other words, let Φ_i^x be the set of features $\varphi \in \Phi_i$ such that $x_{i,\varphi} = x$. Then $c_{\bar{\jmath},x} = \sum_{i \in I, \ \varphi \in \Phi_i^x \cap \Phi'_{i,j_i}} c'_{i,j_i,\varphi}$.

   This sum of functions only concerns features that are both present in $s_i[\ldots]$ and f_i. A feature that is present in $s_i[\ldots]$ but not in f_i, i.e., in $\Phi_i \setminus \Phi_{i,j_i}$, is implicitly considered as having a null cost function. The computation of assignment costs $(c_{\bar{\jmath},x})_{x \in X_r}$ can be implemented using side effects, with a single traversal of the costs functions of alternative $\bar{\jmath}$.

3. Finally, for each cost $(c_{\bar{\jmath},x})_{x \in X_r}$ of an alternative $\bar{\jmath} \in \bar{J}$, which somehow represents an alternative instance of the right-hand side of ρ, we construct the feature assignment cost corresponding to the left-hand side. Given that some variables on the right-hand side (in X_r) are not necessarily present in the left-hand side (in X_0), some feature assignment costs are not propagated at this stage either. As above, they are accumulated in feature \$. Besides, if a variable is only present in the left-hand side, it is useless to propagate the corresponding feature. (Alternatively, it is possible to consider the associated cost function as null.)

   More formally, the feature assignment cost constructed for the left-hand side $s_0[\ldots]$ of ρ, corresponding to alternative $\bar{\jmath}$, is $(c'_{\bar{\jmath},\varphi})_{\varphi \in \Phi'_0}$ where:
   - The propagated features are the feature on the left-hand side, except for the features that are associated to a variable that does not occur on the right-hand side, but including feature \$: $\Phi'_0 = \{\varphi \in \Phi_0 \mid x_{0,\varphi} \notin X_r\} \cup \{\$\}$.
   - For each propagated feature $\varphi \in \Phi'_0$ besides \$, the alternative cost for the corresponding variable $x_{0,\varphi}$ is propagated: $c'_{\bar{\jmath},\varphi} = c_{\bar{\jmath},x_{0,\varphi}}$.
   - Feature \$ accumulates the minimum costs of the feature variables that are not propagated from the right-hand side to the left-hand side: $c'_{\bar{\jmath},\$} = c_{\bar{\jmath},x_\$} + \sum_{x \in X_r \setminus X_0} \min(c_{\bar{\jmath},x})$.

When the roots of the parse forest are reached, a set of alternative costs $(\kappa_i)_{i \in I} = ((c_{i,\varphi})_{\varphi \in \Phi_i})_{i \in I}$ is available. The minimum correction cost is $\min_{i \in I}(\min(\kappa_i))$. This cost may correspond to several alternatives and several feature assignments.

A simple example of such an analysis is shown in figure 2. The global cost for the sentence is 3 ("\$" feature): 2 for the cost of assigning the number feature of "mesure économique" to singular, 1 for the cost of assigning the gender feature of "acceptés" to feminine. (Other features have been discarded for readability.)

Once a feature value assignment with minimum cost is chosen, an opposite top-down traversal (from the roots to the leafs) enumerates actual correction alternatives that implement this minimum cost. More alternatives can be listed

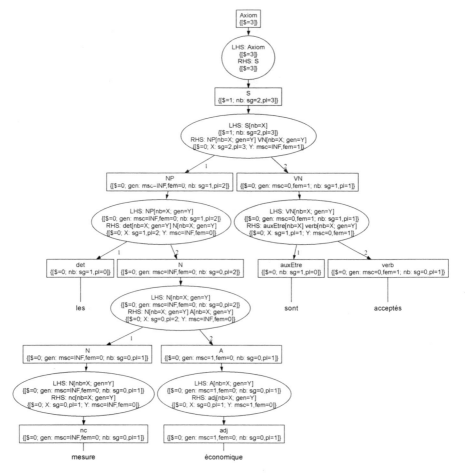

Fig. 2. Correction cost evaluation for "* Les mesure économique sont acceptés"

using cost for ranking. Conversely, if there are too many correction alternatives for a given cost, some may be (randomly) ignored, not to clutter the user with too many choices.

4.2 Combinatorial Reduction and Control

Thanks to the sharing in the parse forest, the algorithm in §4.1 also shares some computations and data structures. Still, it enumerates all parsing alternatives, whose number can be exponential (w.r.t. to the length of the sentence) in the worst case. It is crucial to reduce the number of combinations to consider, without altering the optimum, or using heuristics. For this, we can first set minimum and maximum bounds for a cost function c_φ (cf. example §4.1):

- $\min(c_\varphi) = \min_{u \in Dom(\varphi)}(c_\varphi(u))$ e.g., $\min(c_{\text{genre}}) = \min(0, 1) = 0$
- $\max(c_\varphi) = \max_{u \in Dom(\varphi)}(c_\varphi(u))$ e.g., $\max(c_{\text{genre}}) = \max(0, 1) = 1$

Besides, the minimum and maximum bounds of a feature assignment cost are:

- $\min(\kappa) = \sum_{\varphi \in \Phi} \min(c_\varphi)$ e.g., $\min(\boldsymbol{\kappa}) = \min(0,1) + \min(0,1) = 0$,
- $\max(\kappa) = \sum_{\varphi \in \Phi} \max(c_\varphi)$ e.g., $\max(\boldsymbol{\kappa}) = \max(0,1) + \max(0,1) = 2$,

For any feature assignment α, we are sure that $\min(\kappa) \leq \kappa(\alpha) \leq \max(\kappa)$.

Last, two relations $\leq_\forall$ and $\leq_{\max}^{\min}$ are defined on feature assignment costs. $\leq_\forall$ expresses an upper bound on the costs for each feature value that can be assigned; an absent feature is treated as present with null costs. $\leq_{\max}^{\min}$ expresses the fact that the maximum cost of one assignment is less than the minimum cost of the other one. More precisely, for each $\kappa_1 = (c_{1,\varphi})_{\varphi \in \Phi_1}$ and $\kappa_2 = (c_{2,\varphi})_{\varphi \in \Phi_2}$,

- $\kappa_1 \leq_\forall \kappa_2$ iff $\forall \varphi \in \Phi_1, \forall u \in Dom(\varphi)$, if $\varphi \in \Phi_2$ then $c_{1,\varphi}(u) \leq c_{2,\varphi}(u)$ else $c_{1,\varphi}(u) = 0$,
- $\kappa_1 \leq_{\max}^{\min} \kappa_2$ iff $\max(\kappa_1) \leq \min(\kappa_2)$.

Relation $\leq_\forall$ is a partial order and $\leq_{\max}^{\min}$ is transitive. But, more importantly, they satisfy the following property:

- if $\kappa_1 \leq_\forall \kappa_2$ or $\kappa_1 \leq_{\max}^{\min} \kappa_2$, then for any assignment α: $\kappa_1(\alpha) \leq \kappa_2(\alpha)$.

In this case, if κ_1 and κ_2 are in the same set of alternative costs, whatever the final choice of feature assignment may be, κ_2 will always cost more than κ_1. Cost κ_2 can thus be discarded from the alternatives to consider. This leads to a combinatorial reduction.

To implement this, if $\trianglelefteq$ is one of the relations $\leq_\forall$, $\leq_{\max}^{\min}$, or $\leq_\forall \cup \leq_{\max}^{\min}$ (i.e., the disjunction of both relations $\leq_\forall$ and $\leq_{\max}^{\min}$), we define $\inf_\trianglelefteq$ as the function that takes a set of feature assignment costs $(\kappa_i)_{i \in I}$ and returns its minimal elements with respect to $\trianglelefteq$. In other words, we have:

- $\inf_\trianglelefteq((\kappa_i)_{i \in I}) = (\kappa_j)_{j \in J}$ where $J = \{j \in I \mid \forall i \in I \setminus \{j\}, \kappa_i \ntrianglelefteq \kappa_j\}$

To reduce the number of alternative costs to take into account, without affecting the optimum (cf. §4.1), it is possible to replace a set of alternative feature assignment costs $(\kappa_i)_{i \in I}$ by its minimal elements $\inf_\trianglelefteq((\kappa_i)_{i \in I})$. This applies to:

- case (b): for a forest $f = (t_i)_{i \in I}$, we can reduce using $\inf_\trianglelefteq$ the set union $((c_{i,j,\varphi})_{\varphi \in \Phi_{i,j}})_{i \in I, j \in J_i}$ of the alternative costs corresponding to each tree t_i.
- case (c1): When we rule on features that are not propagated into the right-hand side of ρ, we can $\inf_\trianglelefteq$-reduce the union $((c'_{i,j,\varphi})_{\varphi \in \Phi_{i,j}})_{i \in I, j \in J_i}$ of the alternative costs that, on the contrary, are propagated.
- case (c2): When we construct the agreement cost $(c_x)_{x \in X_r}$ of variables that are in the right-hand side of a rule, we can $\inf_\trianglelefteq$-reduce it, too.
- case (c3): Finally, when we construct the feature assignment cost $(c'_{j,\varphi})_{j \in \bar{J}, \varphi \in \Phi'_0}$ for the left-hand side of a rule, we can also $\inf_\trianglelefteq$-reduce it.

These operations reduce the number of alternative combinations to consider without altering the globality of the optimum.

Although they seem *very* efficient on the kind of sentences and grammars on which we have experimented so far, these optimisations do not guarantee a polynomial time error analysis. To prevent a combinatorial explosion, the following heuristics may also (or alternatively) be used:

- optimistic heuristics: discard any alternative $\kappa = (c_\varphi)_{\varphi \in \Phi}$ such that $\min(\kappa)$ is above a given threshold,
- pessimistic heuristics: discard any alternative $\kappa = (c_\varphi)_{\varphi \in \Phi}$ such that $\max(\kappa)$ is above a given threshold,
- safe heuristics: set an upper bound on the size of any set of alternatives. (In case this upper bound is set to 1, the error analysis computes a local optimum, as opposed to a global one.)

If heuristics are used, the optimum is not global anymore, but the worst case globlal complexity becomes polynomial in the length of the sentence. Only the "safe heuristics" guarantees that a solution is always found.

5   Related Work

Our proposal substantially differs form approaches based on *mal-rules* (aka error productions), where specific, extra rules are added to the parser to treat ill-formed input [13]: ordinary parsing rules are used to detect ill-formed input and mal-rules are meta-rules applying to ordinary rules to relate the structure of ill-formed input to that of well-formed structures, possibly allowing error recovery. With our terminology, ordinary parsing rules form a positive grammar, and mal-rules are specific rules, written in the spirit of a negative grammar, to specify how to relax violated constraints in parsing the positive grammar. By contrast, in our approach, constraint relaxation is automatic, based on feature value accommodation; the linguist does not have to write any specific rule — although he or she has to design the positive grammar to control syntactic proximity.

Mal-rules also result in an explosion of the number of possible parses and raise an issue regarding performance [14]. This can only partially be dealt with using a number of extra rules, many of which are variants of existing rules, which raises another issue regarding grammar maintenance. Although we also use context-free rules augmented with features, we directly address performance with a parser that does enumerate alternatives but construct a shared forest of parse trees. Our cost evaluation also keeps under control the possible combinatorial explosion.

There are proposals to deal with ambiguities using a mal-rule scoring mechanism, the score being defined based on a layered view of stereotypical levels of language ability [15]. Our approach is more compact: it consists in providing a cost model for the plausibility of various feature assignment.

LFG parsing also has been adapted to grammar checking [16]: c-structures are built using a rich error recovery mechanism, and f-structures construction flags unification clashes but always succeeds. Like in our approach, specific rules do not have to be written: constraint relaxation is automatic. Although performance is not mentioned, this could be an issue because LFG grammaticality is NP-complete. Moreover, adding loose error recovery and making all unifications successful dramatically generate ambiguities, which is not mentioned either. We deal with these issues by using context-free rules equiped with flat feature structures, which guaranties a cubic parsing. Although the expressive power may seem poorer, it practically proves to be very flexible and powerful.

The approach in [17] is quite different: parsing amounts to eliminating impractical dependencies using graded constraints (akin to our costs), from an initial state where *all* (exponentially many) dependencies are assumed. A single structural interpretation which violates as few and as weak constraints as possible is picked up by solving a partial constraint satisfaction problem. Remaining dependencies that still violate constraints trigger errors. As for our proposal, the reported error correspond to a global optimum. The design choices regarding performance are quite different though.

6 Conclusion

This paper tackles the central problem of grammar correction and, contrary to common approaches based on negative grammars, argues in favor of a combination of both positive and negative grammars. Moreover, in order to obtain a satisfactory grammar checker, the positive grammars have to provide global and deep analyses supporting feature agreement. We have presented an easy and expressive formalism to precisely do that. The use of a lattice as entry point to our parsing allows the addition of homophones and other "similar" words. Our Earley parser on this lattice relaxes constraints if necessary and provides a shared forest, for which we showed how to compute the most plausible corrections without losing all the combinatorial advantages of the sharing – in a single pass, as opposed to systems such as [18]. We have laid out in greater detail this algorithm, as well as heuristics that control potential combinatorial explosion.

The currently implemented system can, e.g., correct (a) with (b), including the long distance agreement of "cerise" ('cherry') and "cueillis" ('picked'):

Example 4

(a) * Les enfants ont mangé ces cerise rouge qui étaient juteuses et sucrées et qu'ils ont vu que j'avais cueillis.
The children have eaten these-$\langle plur \rangle$ cherry-$\langle fem,sing \rangle$ red-$\langle sing \rangle$ that were-$\langle plur \rangle$ juicy-$\langle fem,plur \rangle$ and sweet-$\langle fem,plur \rangle$ and that they have seen that I have picked-$\langle masc,plur \rangle$.

(b) Les enfants ont mangé ces cerises rouges qui étaient juteuses et sucrées et qu'ils ont vu que j'avais cueillies.
The children have eaten these-$\langle plur \rangle$ cherry-$\langle fem,plur \rangle$ red-$\langle plur \rangle$ that were-$\langle plur \rangle$ juicy-$\langle fem,plur \rangle$ and sweet-$\langle fem,plur \rangle$ and that they have seen that I have picked-$\langle fem,plur \rangle$.

(c) * Les enfants ont mangé cette cerise rouge qui était juteuse et sucrée et qu'ils ont vu que j'avais cueilli.
The children have eaten this-$\langle sing \rangle$ cherry-$\langle fem,sing \rangle$ red-$\langle sing \rangle$ that was-$\langle sing \rangle$ juicy-$\langle fem,sing \rangle$ and sweet-$\langle fem,sing \rangle$ and that they have seen that I have picked-$\langle masc,sing \rangle$.

Only a global and deep analysis can yield this minimal correction (3 revised features). Interestingly, Kleene star and optional terms provide shared forests with few ambiguities. By comparison, Microsoft Office, that often seems to have a somewhat global view of the sentence, actually makes cascading suboptimal choices and corrects (a) into (c), which is not minimal (5 revised features), and is even syntactically wrong as the past participle in the relative phrase is not in agreement. LanguageTool does not pick up any errors in this sentence.

Work is in progress to write a larger grammar and test it on a corpus of errors to assess the quality of error corrections. A crucial extension is likely to be a support for syntactic disambiguation, or more precisely a weighting of alternative parses. We also want to test the system more thoroughly regarding performance and ease in writing grammars. Experiments with other languages are planed too, as well as the integration into OpenOffice.

References

1. Sågvall Hein, A.: A chart-based framework for grammar checking – initial studies. In: 11th Nordic Conference in Computational Linguistic, pp. 68–80 (1998)
2. Souque, A.: Vers une nouvelle approche de la correction grammaticale automatique. In: RECITAL, Avignon, pp. 121–130 (June 2008)
3. Naber, D.: Integrated tools for spelling, style, and grammar checking. In: OpenOffice.org Conference, Barcelona (2007), http://www.languagetool.org.
4. Fontenelle, T.: Les nouveaux outils de correction linguistique de Microsoft. In: TALN Conference, Louvain, pp. 3–19 (April 2006)
5. Clément, L., Sagot, B., Lang, B.: Morphology based automatic acquisition of large-coverage lexica. In: LREC (May 2004),
   http://alpage.inria.fr/~sagot/lefff.html
6. New, B.: Lexique 3: une nouvelle base de données lexicales. In: TALN Conference, pp. 892–900 (April 2006)
7. Prost, J.P.: Modélisation de la gradience syntaxique par analyse relâchée à base de contraintes. Phd thesis, Univ. de Provence et Macquarie Univ. (December 2008)
8. Vogel, C., Cooper, R.: Robust chart parsing with mildly inconsistent feature structures. Edinburh Working Papers in Cognitive Science: Nonclassical Feature Systems 10 (1995)
9. Fouvry, F.: Constraint relaxation with weighted feature structures. In: 8th International Workshop on Parsing Technologies (2003)
10. Billot, S., Lang, B.: The structure of shared forests in ambiguous parsing. In: 27th Meeting on Association for Computational Linguistics, ACL, pp. 143–151 (1989)
11. Earley, J.: An efficient context-free parsing algorithm. CACM 13(2) (1970)
12. Bustamante, F.R., León, F.S.: Gramcheck: A grammar and style checker. In: COLING, pp. 175–181 (1996)
13. Sondheimer, N.K., Weischedel, R.M.: A rule-based approach to ill-formed input. In: 8th Conference on Computational linguistics, Tokyo, Japan, pp. 46–53. ACL (1980)
14. Schneider, D., McCoy, K.F.: Recognizing syntactic errors in the writing of second language learners. In: 17th International Conference on Computational Linguistics, Montreal, Quebec, Canada, pp. 1198–1204. ACL (1998)
15. Mccoy, K.F., Pennington, C.A., Suri, L.Z.: English error correction: A syntactic user model based on principled "mal-rul" scoring. In: 5th International Conference on User Modeling (UM), Hawaii, USA, pp. 59–66 (1996)
16. Reuer, V.: Error recognition and feedback with lexical functional grammar. CALICO Journal 20(3), 497–512 (2003)

17. Menzel, W., Schröder, I.: Constraint-based diagnosis for intelligent language tutoring systems. In: IT&KNOWS Conf. at IFIP 1998, Wien, Budapest, pp. 484–497 (1998)
18. Richardson, S.D., Braden-Harder, L.C.: The experience of developing a large-scale natural language text processing system: CRITIQUE. In: 2nd Conference on Applied Natural Language Processing, Austin, Texas, pp. 195–202. ACL (1988)

D-STAG: A Formalism for Discourse Analysis Based on SDRT and Using Synchronous TAG

Laurence Danlos

Université Paris Diderot, IUF, Alpage
Laurence.Danlos@linguist.jussieu.fr

Abstract. D-STAG is a new formalism for the automatic analysis of discourse. It computes hierarchical discourse structures annotated with discourse relations. They are compatible with those computed in SDRT. A discursive STAG grammar pairs up trees anchored by discourse connectives with trees anchored by (functors associated with) discourse relations.

1 Introduction

The aim of this paper[1] is to propose a new formalism for the automatic analysis of texts, called D-STAG for Discourse Synchronous TAG. This formalism extends a sentential syntactic and semantic analyzer to the discursive level: a discursive analyzer computes the "discourse structure" of the input text. Discourse structures consist of "discourse relations" (also called "rhetorical relations") that link together discourse segments — or more accurately, the meanings these discourse segments convey. A discourse is coherent just in case every proposition that is introduced in the discourse is rhetorically connected to another bit of information, resulting in a connected structure for the whole discourse.

For the discursive part of our analyzer, we rely on SDRT — Segmented Discourse Representation Theory [1,2]. D-STAG computes discourse structures which are compatible with those produced in SDRT. Therefore, D-STAG can take advantage of the results brought by this discourse theory.

The research done in the framework of SDRT is theory-oriented, providing formally detailed accounts of various phenomena pertaining to discourse. Much less focus has been put on the issue of implementing a robust and efficient discourse analyzer. For this aspect of the work, we have designed a formalism based on TAG – Tree Adjoining Grammar [12]. After being used successfully for syntactic analysis in various languages, TAG has been extended in two directions: moving from sentential syntactic analysis to semantic analysis — with, among others, STAG [21,22,17] —, and moving from the sentence level to the discourse level, both for text generation — with, among others, G-TAG [5] —, and for discourse parsing — with, among others, D-LTAG [10]. The new formalism presented here

---

[1] A (longer) French version of this paper is published in *Revue TAL*, Volume 50, Numéro 1, 2009, pp 111-143.

P. de Groote, M. Egg, and L. Kallmeyer (Eds.): FG 2009, LNAI 5591, pp. 64–84, 2011.
© Springer-Verlag Berlin Heidelberg 2011

relies on all this previous work. In particular, its architecture is inspired from that of D-LTAG, with three components:

1. a sentential analyzer, which provides the syntactic and semantic analyses of each sentence in the discourse given as input,
2. a sentence-discourse interface, which is a mandatory component if one wants not to make any change to the sentential analyzer,
3. a discursive analyzer, which computes the discourse structure.

This paper is organized as follows. Section 2 presents the main discursive linguistic data that motivate D-STAG. Section 3 gives an introduction to STAG. Section 4 briefly describes the sentence-discourse interface. Section 5 explains in detail the discursive analyzer. Section 6 compares D-STAG and D-LTAG.

2 Discursive Linguistic Data

A discourse relation is often explicitly expressed by a "discourse connective". The set of discourse connectives includes subordinating and coordinating conjunctions (*because, or*) and discourse adverbials (*next, therefore*). Connectives can be ambiguous. For example, *then* lexicalizes the relation *Narration* in a narrative (*Fred went to the supermarket. Then, he went to the movies.*) whereas it lexicalizes *Continuation* in an enumeration (... *The second chapter presents a state of the art. Then, the third chapter explains the problematics.*). Discourse relations need not be explicitly marked. For example, the relation *Explanation* in the connective-free discourse *Fred fell. Max tripped him up.* of the form $C_0. C_1.$ must be inferred from (extra)linguistic knowledge. For such a case, we assume the existence of an empty adverbial connective, noted ϵ, following a proposition made in [11]. So the previous discourse is assumed to be of the form $C_0. \epsilon C_1.$, and we say, although somewhat inaccurately, that ϵ lexicalizes *Explanation*. In a nutshell, a discourse relation can be considered as a semantic predicate with two arguments which is lexicalized by a discourse connective (possibly empty) with two arguments. The arguments of a discourse relation/connective are the discursive semantic/syntactic representations of the same (continuous) discourse segments. These are the basic principles on which our STAG discursive grammar relies.

One should wonder what discourse structures correspond to when represented as dependency graphs (in which a predicate dominates its arguments). The idea is widespread that dependency graphs representing discourse structures are tree-shaped: this is a basic principle in RST — Rhetorical Structure Theory [14,15] —, a theory on which many text generation or parsing systems have been based for the last twenty years. This is also a principle which guided the conception of D-LTAG. Yet this tree-shaped structure is more a myth than a reality, as shown in [24] and in some of our previous work [6,7]. SDRT discourse structures are not represented as dependency graphs, however, in our two aforementioned papers and in this one, we convert SDRT discourse structures into dependency graphs. These dependency graphs are DAGs — Directed Acyclic Graphs — which are not necessarily tree-shaped. However, these DAGs respect strong constraints which rule out a number of DAGs that don't correspond to any discourse structure. We shall justify this claim with discourses of the form C_0 *because* C_1. Adv_2 C_2., in which *because* lexicalizes *Explanation*; C_i symbolizes the ith clause, its logical form is noted F_i. These discourses yield four types of interpretation — but no more than four — which are illustrated in examples (1).[2]

(1) a. Fred is in a bad mood because he lost his keys. Moreover, he failed his exam.

   b. Fred is in a bad mood because he didn't sleep well. He had nightmares.

   c. Fred went to the supermarket because his fridge is empty. Then, he went to the movies.

   d. Fred is upset because his wife is abroad for a week. This shows that he does love her.

In (1a), $Adv_2 = moreover$ lexicalizes the relation *Continuation*. The discourse segment C_1. Adv_1 C_2 forms a complex constituent whose logical form, $Continuation(F_1, F_2)$, is the second argument of *Explanation*. So the discourse structure is $Explanation(F_0, Continuation(F_1, F_2))$, which corresponds to a tree-shaped dependency DAG, see the adjacent figure with $R_1 = Explanation$ and $R_2 = Continuation$. In this example, the second discursive argument of the conjunction *because* crosses a sentence boundary.

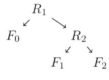

   In (1b), $Adv_2 = \epsilon$ lexicalizes *Explanation*. The discourse structure is $Explanation(F_0, F_1) \wedge Explanation(F_1, F_2)$, which corresponds to a non tree-shaped dependency DAG.

   In (1c), $Adv_2 = then$ lexicalizes *Narration*. The discourse structure is $Explanation(F_0, F_1) \wedge Narration(F_0, F_2)$, which corresponds to a non tree-shaped dependency DAG.

   In (1d), $Adv_2 = \epsilon$ lexicalizes *Commentary*. The discourse segment C_0 *because* C_1 forms a complex constituent whose logical form is the first argument of *Commentary*. So the discourse structure is $Commentary(Explanation(F_0, F_1), F_2)$, which corresponds to a tree-shaped dependency DAG.

In conclusion, these empirical data show that a formalism for the automatic analysis of discourse must be able to compute dependency structures which are not

---

[2] In [13], these four types of interpretation are finally brought to light, thanks to the Penn Discourse Tree Bank (PDTB), which is an English corpus manually annotated for discourse relations and their arguments [18].

tree-shaped.[3] This principle guided the conception of D-STAG. More precisely, from this (and other) data, we laid down the constraints below which govern the arguments of a discourse connective/relation using the following terminology. The clause in which a connective appears is called its "host clause". An adverbial connective appears in front of its host clause or within its VP. A subordinating conjunction always appears in front of its host clause, which is called an "adverbial clause." At the sentence level, an adverbial clause modifies a "matrix clause." It is located on its right, on its left, or inside it before its VP. When it is located on its right, the subordinating conjunction is said to be "postposed," otherwise it is said to be "preposed." A discourse connective/relation has two arguments which are the syntactic/semantic representations of two discourse segments that we call the "host segment" and the "mate segment". These segments are governed by the following constraints.

Constraint 1. *The host segment of a connective is identical to or starts at its host clause (possibly crossing a sentence boundary).*

Constraint 2. *The mate segment of an adverbial is anywhere on the left of its host segment (generally crossing a sentence boundary).*[4]

Constraint 3. *The mate segment of a postposed conjunction is on the left of its host segment without crossing a sentence boundary.*

Constraint 4. *The mate segment of a preposed conjunction is identical to or starts at the matrix clause (possibly crossing a sentence boundary).*

3 Introduction to TAG and STAG

This section is reproduced except where noted from [17] with permission of the authors. It begins with a brief introduction to the use of TAG in syntax.

"A tree-adjoining grammar (TAG) consists of a set of elementary tree structures and two operations, substitution and adjunction, used to combine these structures. The elementary trees can be of arbitrary depth. Each internal node is labeled with a nonterminal symbol. Frontier nodes may be labeled with either terminal symbols or nonterminal symbols and one of the diacritics $\downarrow$ or $*$. Use of the diacritic $\downarrow$ on a frontier node indicates that it is a *substitution node*. The *substitution* operation occurs when an elementary tree rooted in the nonterminal symbol A is substituted for a substitution node labeled with the nonterminal symbol A. Auxiliary trees are elementary trees in which the root and a frontier node, called the *foot node* and distinguished by the diacritic $*$, are labeled with the same nonterminal. The *adjunction* operation involves splicing an auxiliary

---

[3] In RST, (1b) and (1c) are represented as trees that must be interpreted with the "Nuclearity Principle" [16]. However, as explained in [7], the Nuclearity Principle leads to a wrong interpretation for (1d), namely $Explanation(F_0, F_1) \wedge Commentary(F_0, F_2)$.

[4] However, the mate segment must conform to the Right Frontier Constraint, which has been postulated in SDRT, see Sect. 5.1.

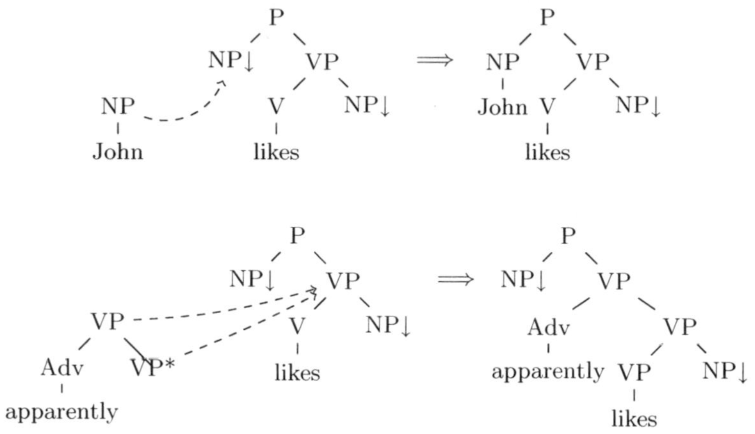

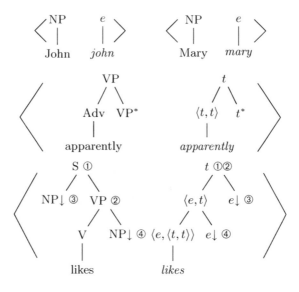

Fig. 1. Example TAG substitution and adjunction operations (reproduced from [17])

Fig. 2. An English syntax/semantics STAG fragment (reproduced from [17])

tree with root and designated foot node labeled with a nonterminal A at a node
in an elementary tree also labeled with nonterminal A. Examples of the substitution and adjunction operations on sample elementary trees are shown in
Figure 1."

"Synchronous TAG (STAG) extends TAG by taking the elementary structures
to be pairs of TAG trees with links between particular nodes in those trees. An
STAG is a set of triples, $\langle t_L, t_R, \frown \rangle$ where t_L and t_R are elementary TAG trees and
$\frown$ is a linking relation between nodes in t_L and nodes in t_R [21,22]. Derivation

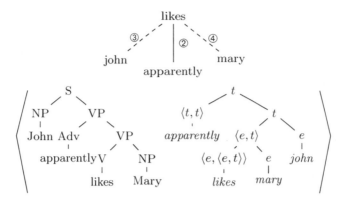

Fig. 3. Derivation tree and derived tree pair for *John apparently likes Mary* (reproduced from [17])

proceeds as in TAG except that all operations must be paired. That is, a tree can only be substituted or adjoined at a node if its pair is simultaneously substituted or adjoined at a linked node." We notate the links by using circled indices (e.g. ①) marking linked nodes.

STAG has been successfully used in an English sentential syntax/semantics interface [17]. For the sentence *John apparently likes Mary*, Fig. 2 gives the STAG fragment, Fig. 3 the derivation tree and the derived tree pair. In derivation trees, "substitutions are notated with a solid line and adjunctions are notated with a dashed line. Note that each link in the derivation tree specifies a link number in the elementary tree pair. The links provide the location of the operations in the syntax tree and in the semantics tree. These operations must occur at linked nodes in the target elementary tree pair. In this case, the noun phrases *John* and *Mary* substitute into *likes* at links ③ and ④ respectively. The word *apparently* adjoins at link ②. The resulting semantic representation can be read off the derived tree by treating the leftmost child of a node as a functor and its siblings as its arguments. Our sample sentence thus results in the semantic representation *apparently(likes(john, mary))*."

4   Sentence-Discourse Interface

We first explain why this interface is necessary. The idea in D-STAG is to extend a sentential analyzer to the discourse level **without making any change to it**. Yet, one cannot directly pass from sentence to discourse because there are mismatches between the arguments of a connective at the discourse level and its arguments at the sentence level. First, an adverbial connective has compulsorily two arguments at the discourse level, whereas it has only one argument at the sentence level. Second, a subordinating conjunction can have an argument at the discourse level which crosses a sentence boundary (see (1a) and (3) below), whereas this is out of the question at the sentence level.

In conclusion, it is necessary to pass through a sentence-discourse interface which gives sentence boundaries the simple role of punctuation signs and which allows us to re-compute the (two) arguments of a connective. Such an interface is also used in D-LTAG, by which we were inspired. From the sentential syntactic analysis, this interface deterministically produces a "Discourse Normalized Form" (henceforth DNF), which is a sequence of "discourse words:" a discourse word is mainly a connective, an identifier C_i for a clause (without any connective) or a punctuation sign. The syntactic and semantic analyses for C_is are those obtained by the sentential analyzer by removing connectives. An adverbial connective is moved in front of its host clause if not already there, while keeping a trace of its original position. If a normalized sentence (except the very first one) doesn't start with an adverbial connective, the empty connective ϵ is introduced. As an illustration, for (2), the DNF is C_0. ϵ as C_1, C_2. $then^{internal}$ C_3 because C_4.

(2) Fred went to the movies. As he was in a bad mood, he didn't enjoy it. He then went to a bar because he was dead thirsty.

The sequence of discourse words making up a DNF follows a regular grammar. In Section 5.1 dedicated to adverbial connectives and postposed conjunctions, a DNF follows the regular expression C $(Punct\ Conn\ C)^*$, in which the sequence $Punct\ Conn$ is either . Adv or $(,)$ $Conj$ where the comma is optional. Disregarding punctuation signs, the DNF is: $C_0\ Conn_1\ C_1 \ldots Conn_n\ C_n$, with $Conn_i = Adv_i$ ou $Conj_i$. A DNF with a preposed conjunction (Sect. 5.2) includes one element C which is preceded by the expressions $Conj\ C((,)\ Conj\ C)^*$. Connectives can be optionally followed or preceded by a modifier (Sect. 5.3). Coordinating conjunctions are studied in Section 5.4. Cases of "multiple connectives" in which two connectives share the same host clause are studied in Section 6.

This regular grammar doesn't decompose clauses into sub-clauses: it takes into account neither clausal complements nor incident clauses nor relative clauses, while these sub-clauses may play a role at the discourse level. We plan to complete the regular grammar for DNFs in future research and to extend the discursive component of D-STAG accordingly (Sect 7).

5   Discursive Component of D-STAG

For a clause C_i (without any connective), the sentential-discourse interface provides its syntactic tree rooted in S and noted T_i, its semantic tree rooted in t and noted F_i, and its derivation tree noted τ_i. To plug the clausal analyses into the discourse ones, we use the pair $\alpha S\text{-}to\text{-}D$ given in Fig. 4-a, in which the symbol DU represents the category "discourse unit". In the rest of this paper, we note η_i the derivation tree made of $\alpha S\text{-}to\text{-}D$ in which τ_i is substituted at link $\odot$; η_i corresponds to the pair given in Fig. 4-b. We also use the following convention: as any tree of our grammar includes at the most one substitution node, this one (when it exists) is systematically marked with link $\odot$.

When a given connective $Conn_i$ lexicalizes a single discourse relation R_i, the basic principle of the discursive STAG grammar consists in designing a tree pair,

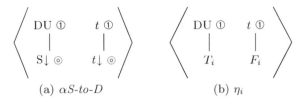

(a) αS-to-D (b) η_i

Fig. 4. Tree pairs αS-to-D and η_i

noted $Conn_i \div R_i$, whose syntactic tree is anchored by $Conn_i$ and whose semantic tree is anchored by a lambda-term associated with R_i. We say, although
somewhat inaccurately, that the semantic tree is anchored by R_i. When a connective is ambiguous, i.e. it lexicalizes several discourse relations, it anchors as
many syntactic trees as discourse relations it lexicalizes (this is in particular the
case for the empty connective ϵ). However, ambiguity issues are not in the scope
of this paper.

We start with the presentation of the STAG discursive grammar for adverbials
and postposed conjunctions, which are connectives with a similar behavior.

5.1   Adverbial Connectives and Postposed Conjunctions

Syntactic trees. The syntactic trees anchored by an adverbial connective or by
a postposed conjunction are given in Fig. 5, in which a discourse connective is
of category DC. Disregarding the features for now, these trees differ only in the
co-anchors which are punctuation signs of category Punct.

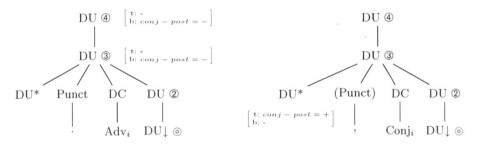

Fig. 5. Syntactic trees for adverbial connectives and postposed conjunctions

These trees observe the following principles: they are **auxiliary** trees with
two arguments given by a substitution node DU↓ and a foot node DU*. The
substitution node is for the host argument of the connective, i.e. the DU for the
host segment substitutes at DU↓. If the DU for the host clause has undergone
an adjunction, then the host segment starts at — but is not identical to —
the host clause, see Constraint 1 in Sect. 2. The foot node corresponds to the
mate argument of the connective. It is located on its left, which conforms to
Constraints 2 and 3. The fact that the mate segment of a postposed conjunction

cannot cross a sentence boundary, contrarily to that of an adverbial, is handled with features $[conj - post = \pm]$ explained later.

We postulate an incremental analysis procedure in which the sequence of the discourse words of a DNF of the form C_0 $Conn_1$ C_1 ... $Conn_n$ C_n (disregarding punctuation signs) is analyzed from left to right. After analyzing $C_0 \ldots Conn_n$ C_n, the attachment of the new connective $Conn_{n+1}$ is realized by **adjunction** of the tree anchored by $Conn_{n+1}$ at a node DU on the **right frontier** of the syntactic tree representing $C_0 \ldots Conn_n$ C_n. The attachment of the new clause C_{n+1} is realized by substituting the syntactic tree of the pair η_{n+1} at the substitution node DU↓ of the tree anchored by $Conn_{n+1}$. Let us underline that this incremental procedure doesn't take into account the segmentation of the discourse into sentences.

Trees anchored by an adverbial or a postposed conjunction include three nodes labelled DU, with link ②, ③ or ④, on their right frontier. These nodes are marked with different links, which allows us to get various semantic interpretations, as shown below. There exist three nodes DU with different links, and not a single node DU with three different links, so as to allow several adjunctions to different nodes, for example an adjunction at DU③ to attach $Conn_n$ and an adjunction at DU④ to attach $Conn_{n+1}$. It should be noted that if an adjunction is done at DU③ to attach $Conn_n$, DU② is no longer on the right frontier of the syntactic tree. Therefore, DU② can no longer be an adjunction site to attach $Conn_{n+1}$. This constraint will be generalized in Constraint 5 below.

Semantic trees. At first sight, one could consider that a discourse relation R_i is associated with the functor $\mathcal{R}_i = \lambda xy.R_i(x,y)$ with $x, y : t$, $R_i(x,y) : t$, and $\mathcal{R}_i : \langle t, \langle t, t \rangle \rangle$, $\mathcal{R}_i$ anchoring a tree with a foot node $t*$ and a substitution node $t\downarrow$. Yet this is appropriate only to analyze a simple DNF with two clauses, for example a DNF of the form C_0 *because* C_1 as shown in Fig. 6 in which $\beta_1 = because_{post} \div Explanation$.

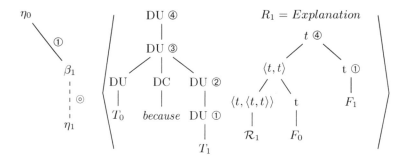

Fig. 6. Derivation tree and derived tree pair for a DNF of the form C_0 *because* C_1 (using the functor $\mathcal{R}_1$ in the semantic tree)

However, with this simple functor $\mathcal{R}_i$, it is impossible to obtain four interpretations (of which two are a conjunction of formulae) for DNFs with three clauses (see Sect. 2). Therefore, we define two type-shifting operators Φ' and Φ'': they take $\mathcal{R}_i$ as argument and return two new functors $\mathcal{R}'_i$ and $\mathcal{R}''_i$ associated with the discourse relation R_i. Below the definitions.

Definition 1. $\Phi' = \lambda\mathcal{R}_i XY.X(\lambda x.Y(\lambda y.\mathcal{R}_i(x,y)))$
$\Phi'(\mathcal{R}_i) = \mathcal{R}'_i = \lambda XY.X(\lambda x.Y(\lambda y.\mathcal{R}_i(x,y)))$
with $X, Y : ttt = \langle\langle t,t\rangle, t\rangle$ and $x, y : t$

Φ' triggers a type raising on the arguments. The resulting functor $\mathcal{R}'_i$ is of type $\langle ttt, \langle ttt, t\rangle\rangle$ in which ttt symbolizes the type $\langle\langle t,t\rangle, t\rangle$. It co-anchors tree (A), given in Fig. 7-a, whose foot node is of type t. (A) is used for adjunctions at links ① and ④. If the first argument of $\mathcal{R}'_i$ is $\lambda P.P(F_0)$ of type ttt, the second one $\lambda Q.Q(F_1)$ of type ttt, then the result is $R_i(F_0, F_1)$ of type t. So, for a DNF with two clauses, $\mathcal{R}'_i$ leads to the same result as $\mathcal{R}_i$. Yet, the type raising is necessary to introduce nodes ttt ② and ttt ③ at which (B) can adjoin.

Definition 2. $\Phi'' = \lambda\mathcal{R}_i XYP.X(\lambda x.Y(\lambda y.\mathcal{R}_i(x,y) \wedge P(x)))$
$\Phi''(\mathcal{R}_i) = \mathcal{R}''_i = \lambda XYP.X(\lambda x.Y(\lambda y.\mathcal{R}_i(x,y) \wedge P(x)))$
with $X, Y : ttt = \langle\langle t,t\rangle, t\rangle$, $P : \langle t,t\rangle$ and $x, y : t$

Φ'' introduces a conjunction of terms. The resulting functor $\mathcal{R}''_i$ is of type $\langle ttt, \langle ttt, ttt\rangle\rangle$. It anchors tree (B), given in Fig. 7-b, whose foot node is of type ttt. (B) is used for adjunctions at links ② and ③. If the first argument of $\mathcal{R}''_i$ is $\lambda P.P(F_0)$, the second one $\lambda Q.Q(F_1)$, then the result is $\lambda P.(R_i(F_0, F_1) \wedge P(F_0))$ of type ttt.

Analysis of DNFs with three clauses. For DNFs with three clauses, four types of interpretation must be computed. These were illustrated in examples (1) of the form C_0 *because* C_1. *Adv$_2$* C_2 in Sect. 2, and we are going to explain the analysis of these examples. We call β_1 the tree pair *because$_{post}$* $\div$ *Explanation* and β_2 the pair *Adv$_2$* $\div R_2$. After analyzing C_0 *because* C_1, the syntactic tree is that shown in Fig. 6. The right frontier of this tree includes four nodes labelled DU which can receive the adjunction of the syntactic tree of β_2. These nodes are marked with link ① coming from the syntactic tree of η_1 or link ②, ③ or ④ coming from the syntactic tree anchored by *because*. The analyses of the four examples in (1) are obtained by adjoining β_2 at one of these links.

We start with (1a) with $\beta_2 = $ *moreover* $\div$ *Continuation*, for which the discourse structure is $Explanation(F_0, Continuation(F_1, F_2))$. This is obtained by adjoining β_2 at link ① of η_1. The node with link ① in the semantic tree of η_1 is of type t. Therefore, one must use tree (A) anchored by $\mathcal{R}'_2$, whose foot node is of type t. The semantic derived tree for (1a) is given in Fig. 8. The sub-tree rooted at Gorn address 2 results in $\lambda P.P(Continuation(F_1, F_2))$ with $P : \langle t,t\rangle$. So $Continuation(F_1, F_2)$ is the second argument of $R_1 = Explanation$, whose first argument is F_0, hence the formula $Explanation(F_0, Continuation(F_1, F_2))$.

We go on with (1b) with $\beta_2 = \epsilon \div$ *Explanation*, for which the discourse structure is $Explanation(F_0, F_1) \wedge Explanation(F_1, F_2)$ with a conjunction of formulae. This is obtained by adjoining β_2 at link ② of β_1. The node with link ② in the semantic tree of β_1 is of type ttt. Therefore, one must use tree

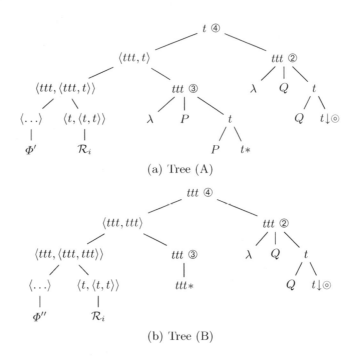

(a) Tree (A)

(b) Tree (B)

Fig. 7. Semantic trees (A) and (B) anchored by $\mathcal{R}'_i = \phi'(\mathcal{R}_i)$ and $\mathcal{R}''_i = \phi''(\mathcal{R}_i)$

(B) anchored by $\mathcal{R}''_2$, whose foot node is of type ttt. The semantic derived tree for (1b) is given in Fig. 9-a. The sub-tree rooted at Gorn address 2 results in $\lambda P.(Explanation(F_1, F_2) \wedge P(F_1))$ with $P : \langle t, t \rangle$. As only F_1 is under P, it is the second argument of $R_1 = Explanation$, whose first argument is F_0, hence the formula $Explanation(F_0, F_1) \wedge Explanation(F_1, F_2)$.

For (1c) with $\beta_2 = then \div Narration$, the structure is $Explanation(F_0, F_1) \wedge Narration(F_0, F_2)$ with also a conjunction of formulae. This is obtained by adjoining β_2 at link ③ of β_1. This case is similar to the previous one, so we simply give the semantic derived tree in Fig. 9-b.

Let us finish with (1d) with $\beta_2 = \epsilon \div Commentary$, for which the structure is $Commentary(Explanation(F_0, F_1), F_2)$. This is obtained by adjoining β_2 at link ④ of β_1. The node with link ④ in the semantic tree of β_1 is of type t. Therefore, one must use tree (A) anchored by $\mathcal{R}'_2$. The semantic derived tree for (1d) is given in Fig. 10. The sub-tree rooted at Gorn address 1.2 results in $\lambda P.P(Explanation(F_0, F_1))$ with $P : \langle t, t \rangle$.

In conclusion, the four types of interpretation of DNFs of the form $C_0 \, Conj_1 \, C_1$. $Adv_2 \, C_2$ are computed thanks to the four adjunction sites on the right frontier of the syntactic tree for $C_0 \, Conj_1 \, C_1$ and to semantic trees (A) and (B) whose foot nodes are respectively of type t et ttt and which are anchored by $\mathcal{R}'_i$ and $\mathcal{R}''_i$.

For DNFs with three clauses of the form $C_0 \, Conn_1 \, C_1 \, Conn_2 \, C_2$, we have just examined the case $C_0 \, Conj_1 \, C_1$. $Adv_2 \, C_2$ where $Conn_1$ is a postposed conjunction and $Conn_2$ an adverbial. Three cases are left: $Conn_1$ is a postposed

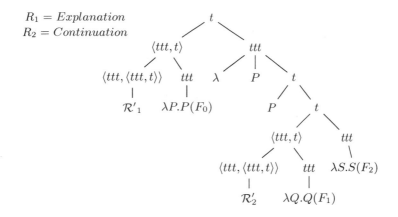

Fig. 8. Semantic derived tree for (1a) with interpretation $R_1(F_0, R_2(F_1, F_2))$

conjunction and $Conn_2$ also, $Conn_1$ is an adverbial and $Conn_2$ also, $Conn_1$ is an adverbial and $Conn_2$ a postposed conjunction. The first two cases raise no new issues. The third case, which concerns DNFs of the form $C_0.\ Adv_1\ C_1\ Conj_2\ C_2$, raises the issue of fully implementing Constraint 3 from Sect. 2, which states that the mate segment of a postposed conjunction cannot cross a sentence boundary. This constraint is implemented thanks to features $[conj - post = \pm]$ which decorate some nodes in the syntactic trees anchored by an adverbial or a postposed conjunction, which are given in Fig. 5. More precisely:

- the foot node of a tree anchored by a postposed conjunction is decorated with the top feature $conj - post = +$,
- the nodes with links ③ and ④ in a tree anchored by an adverbial are decorated with the bottom feature $conj - post = -$.

These features block the adjunction of $Conj_2 \div R_2$ at links ③ and ④ of $Adv_1 \div R_1$ thanks to unification failure $(conj - post = +) \cup (conj - post = -)$. Therefore, $Conj_2 \div R_2$ can only adjoin at link ② of $Adv_1 \div R_1$ and at link ① of η_1, which results respectively in interpretations $R_1(F_0, F_1) \wedge R_2(F_1, F_2)$ and $R_1(F_0, R_2(F_1, F_2))$. These interpretations conform to Constraint 3: the mate argument of R_2 is F_1, so the mate segment of $Conj_2$ is C_1, which doesn't cross a sentence boundary.

Analysis of DNFs *with n clauses $(n > 3)$ of the form* $C_0\ Conn_0\ C_1\ \ldots\ C_n$. For DNFs with n clauses, no new mechanism is involved. Attaching $Conn_{n+1}$ next C_{n+1} consists, in the derivation tree representing $C_0 \ldots C_n$, in adjoining $Conn_{n+1} \div R_{n+1}$ — in which η_{n+1} is substituted — at link ① of η_n or at link ⓘ with $i \in \{2, 3, 4\}$ of a node $\beta_k = Conn_k \div R_k$, the node at link ⓘ coming from the syntactic tree of β_k being on the right frontier of the syntactic tree for $C_0 \ldots C_n$ (to keep to the linear order of the DNF). As it is long and tedious

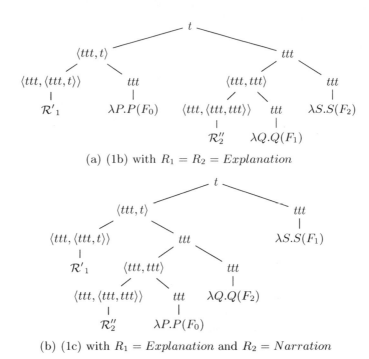

(a) (1b) with $R_1 = R_2 = Explanation$

(b) (1c) with $R_1 = Explanation$ and $R_2 = Narration$

Fig. 9. Semantic derived trees for (1b) and (1c) with interpretations $R_1(F_0, F_1) \wedge R_2(F_1, F_2)$ and $R_1(F_0, F_1) \wedge R_2(F_0, F_2)$

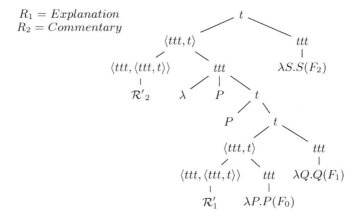

Fig. 10. Semantic derived trees for (1d) with interpretation $R_2(R_1(F_0, F_1), F_2)$

to determine (the right frontier of) the derived syntactic tree, it is convenient to define a notion of right frontier on the derivation tree. Since derivation trees are intrinsically not ordered, a graphical convention which represents derivation trees as ordered must be called upon. The following convention is satisfactory:

the nodes labeled η_k projected onto a line are ordered by following the linear order of the DNF. With this order relation noted $\prec$, the nodes β_k in the derivation tree which are possible sites of adjunction for a new segment are those on the right frontier of the derivation tree, while observing Constraint 5 which governs two adjunctions at links $\textcircled{n}$ and $\textcircled{m}$ of the same node:[5]

Constraint 5. *If β_j — in which η_j is substituted — adjoins at link $\textcircled{n}$ of a node β_i, then β_k — in which η_k is substituted — can adjoin at link $\textcircled{m}$ of the same node β_i only if the following rule is observed: $\eta_j \prec \eta_k \Rightarrow n < m$ (with n and m belonging to $\{2, 3, 4\}$).*

This constraint generalizes the one we formulated before, namely, if an adjunction is performed at node DU③ of a syntactic tree anchored by $Conn_i$, then a new adjunction in this tree can be performed at DU④ but not at DU②.

Implementation of RFC. The Right Frontier Constraint (RFC) postulated in SDRT relies on a distinction between two types of discourse relations: coordinating (*Narration, Continuation*) versus subordinating (*Explanation, Commentary*) ones. RFC says that it is forbidden to attach new information to the first argument of a coordinating relation [2]. As a consequence, for example, the interpretation $R_1(F_0, F_1) \wedge R_2(F_0, F_2)$ is excluded when R_1 is coordinating.

Implementing RFC in D-STAG requires first to distinguish the semantic trees anchored by a coordinating versus subordinating relation. This is achieved by creating two copies of semantic trees (A) and (B), which differ in a top feature $[coord = \pm]$ decorating their foot node. Then RFC is implemented by forbidding any adjunction at link ③ of the copies of (A) and (B) whose foot node is decorated with the feature $[coord = +]$: it is link ③ which is used to obtain interpretations like $R_1(F_0, F_1) \wedge R_2(F_0, F_2)$ which must be excluded when R_1 is coordinating.

SDRT postulates other semantic constraints based on the distinction between coordinating versus subordinating relations, for example the "Continuous Discourse Pattern" [3]. There is no room to describe them, however we can say that they are easily implemented in D-STAG thanks to the addition of a set of features in the semantic trees.

5.2   Preposed Conjunctions

The syntactic tree anchored by a preposed conjunction is given in Fig. 11. It is designed so as to respect Constraints 1 and 4 given in Sect. 2. It differs from the syntactic trees anchored by an adverbial or a postposed conjunction, among other things, by the fact that the foot node DU∗ is dominated by a node DU with link ⑤. To take into account this new link, a node $t⑤$ dominating the foot node $t∗$ is added into the two copies of semantic tree (A).

An adjunction to link ⑤ is used when the preposed conjunction introduces a "framing adverbial" [4] as illustrated in (3) of the form $When\ C_0, C_1.\ Next\ C_2$. In

---

[5] This constraint is valid because we assigned links ②, ③ and ④ in a well-thought-out manner.

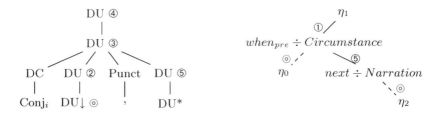

Fig. 11. Tree anchored by a preposed conjunction and derivation tree for (3)

this discourse, the preposed conjunction *when* has its mate segment which crosses a sentence boundary. Its interpretation is $Circumstance(Narration(F_1, F_2), F_0)$, assuming that *when* conveys *Circumstance* and *next* conveys *Narration*.

(3) When he was in Paris, Fred went to the Eiffel Tower. Next, he visited The Louvre.

The derivation tree for (3) is given in Fig. 11. The semantic trees anchored by *Circumstance* and *Narration* are both (A). The arguments of the functor associated to *Circumstance* are $\lambda P.P(Narration(F_1, F_2))$ and $\lambda Q.Q(F_0)$. Therefore, $Narration(F_1, F_2)$ is the first argument of *Circumstance*, F_0 the second one.[6]

5.3   Modifiers of Discourse Connectives/Relations

As far as we are aware, modification of discourse connectives/relations is a phenomenon which has been neglected. However, it is a common phenomenon as illustrated in (4).

(4)   a. Fred is in a bad mood *only/even/except* when it is sunny.
    b. You shouldn't trust John because, *for example*, he never returns what he borrows. [23]
    c. John just broke his arm. So, *for example*, he can't cycle to work.[23].

In [23], *for example* is not considered as a connective modifier but as a connective whose interpretation is "parasitic" on the relation conveyed by the connective on its left. This position, which is not linguistically justified, leads to laborious computations in D-LTAG [10, pp 31-35] to obtain the interpretation of (4b). On the contrary, in D-STAG, we propose that *for example* in (4b) or (4c) is a modifier of the connective on its left in the same way that *only, even* and *except* in (4a) are modifiers of the connective on their right. This position sounds more justified on linguistic grounds and it allows us to obtain the interpretation of a discourse such as (4b) in a very simple way, as we are going to show.

---

[6] Examples such as (3) make it that, despite appearances, the syntactic discursive grammar of D-STAG is not a TIG (Tree Insertion Grammar) [20], since the right tree anchored by *next* is adjoined on the spine of the left tree anchored by the preposed conjunction *when*.

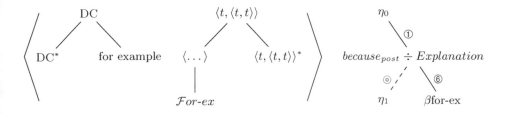

Fig. 12. Pair $\beta$for-ex and derivation tree for (4b)

In D-STAG, connective modifiers anchor (syntactic) auxiliary trees whose foot node is DC (left tree for *only, even* and *except*, right tree for *for example*). To adjoin these trees, we mark the node labelled DC with link ⑥ in the syntactic trees for connectives (Fig. 5 and 11). At the semantic level, we assume that the contribution of a discourse relation modifier consists in transforming a functor $\mathcal{R}_i$ of type $\langle t, \langle t, t \rangle \rangle$ into another functor of the same type. Therefore, the nodes dominating $\mathcal{R}_i$ are marked with link ⑥ in the two copies of (A) and (B). Let us illustrate adjunctions to link ⑥ with (4b) of the form C_0 *because for example* C_1. As explained in [23], the interpretation of (4b) is $Exemplification(F_1, \lambda r.Explanation(F_0, r))$ with $r : t$. To get this interpretation, we define the functor $\mathcal{F}or\text{-}ex$ as below. The pair named $\beta$for-ex is given in Fig. 12, which also shows the derivation tree for (4b). The functor $\Phi'(\mathcal{F}or\text{-}ex\,(\mathcal{R}_i))$ with $R_i = Explanation$ results in the right interpretation.

Definition 3. $\mathcal{F}or\text{-}ex = \lambda \mathcal{R}_i pq.Exemplification(q, \lambda r.\mathcal{R}_i(p, r))$
with $\mathcal{R}_i : \langle t, \langle t, t \rangle \rangle$ and $p, q : t$.

5.4   Coordinating Conjunctions

For reasons that will be obvious later, we start with correlative coordinations of adverbial clauses, which are illustrated in (5). These correlative coordinations can easily be handled in D-STAG by considering e.g. *neither* and *nor* as (adverbial) modifiers of the subordinating conjunction on their right. The correlative part of these constructions, e.g. the fact that *neither* cannot modify a subordinating conjunction without the presence of *either* modifying another subordinating conjunction, is taken into account by a set of features in the syntactic trees anchored by a subordinating conjunction (there is no room to explain these features). This set of features also forces the adjunction of the tree anchored by the second (modified) subordinating conjunction to link ③ of the tree anchored by the first (modified) subordinating conjunction. As a result, the interpretation is a conjunction of formulae in which the logical form of the matrix clause is shared. More precisely, for (5a) of the form C_0 *neither when* C_1 *nor when* C_2, the interpretation $\neg Condition(F_0, F_1) \wedge \neg Condition(F_0, F_2)$ — assuming that *when* lexicalizes the relation $Condition$ — is obtained (through an adjunction of the second *when* at link ③ of the first *when*) by giving *neither* and *nor* the semantics of negation. For (5b) of the form C_0 *either if* C_1 *or if* C_2, the interpretation $Condition(F_0, F_1) \vee Condition(F_0, F_2) = \neg(\neg Condition(F_0, F_1) \wedge \neg Condition(F_0, F_2))$ is obtained by giving to *or* the semantics of negation and by associating to *either* a semantic tree with two parts, one for the local scope of negation, the other one for the global scope of negation over the conjunction of formulae.[7] For (5c) of the form C_0 *both when* C_1 *and when* C_2, the interpretation $Condition(F_0, F_1) \wedge Condition(F_0, F_2)$ is obtained by giving to *both* and *and* the semantics of identity.

(5) a. Fred is pleased *neither* when it is sunny *nor* when it is rainy.
  b. Fred will come *either* if it is sunny *or* if it is rainy.
  c. Fred is pleased *both* when it is sunny *and* when it is rainy.

We now examine non-correlative coordinations. Coordinations of two matrix clauses with a DNF of the form C_0 $Coord_1$ C_1 are handled with a syntactic tree anchored by the coordinating conjunction $Coord_1$ similar to the syntactic tree anchored by a subordinating conjunction (Fig. 5). Coordination of n clauses with $n > 2$ with a DNF of the form C_0, $C_1(,)$ $Coord_2$ C_2 are handled with the standard mechanisms for multiple coordinations. Coordinations of adverbial clauses with a DNF of the form C_0 $Conj_1$ C_1 *or/and* $Conj_2$ C_2 raise drastic problems, in particular to get their interpretation. However, it must be noted that the interpretation of (5b)/(5c) doesn't change if *either/both* is omitted. In other words, a non-correlative coordination of adverbial clauses is semantically equivalent to a correlative coordination. Therefore, we propose that the sentence-discourse interface automatically converts a non-correlative coordination of adverbial clauses into a correlative one, e.g. converts the DNF C_0 $Conj_1$ C_1 *or* $Conj_2$ C_2 into the DNF C_0 *either* $Conj_1$ C_1 *or* $Conj_2$ C_2 that we can handle by considering *either* and *or* as modifiers of the subordinating conjunction on their right.

6   Comparison between D-STAG and D-LTAG

D-STAG and D-LTAG [10] roughly share the same goal and the same architecture. The discrepancies between these two formalisms mainly lie in the discursive component.[8] First, D-LTAG makes little use of discourse relations and ignores the distinction between coordinating versus subordinating relations. In short, it doesn't build on discourse theories. This is even a principle, as shown in the quotation [10, p 1] " D-LTAG presents a model of low-level discourse structure and interpretation that exploits the *same* mechanisms used at the sentence level and builds them directly on top of clause structure and interpretation". This prevents D-LTAG from taking advantage of the insights provided by discourse theories, which supply rhetorical knowledge, among other things.

---

[7] This is the only case which requires a multi-component semantic tree.
[8] The sentential components of D-STAG and D-LTAG are also crucially different, since D-STAG relies on STAG, which is not the case for D-LTAG. However, the sentential level is out of the scope of this paper, see [17] for a discussion on the various approaches for a sentential syntax-semantics interface.

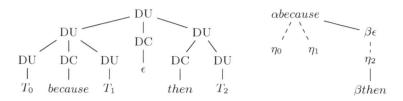

Fig. 13. D-LTAG syntactic tree and derivation tree for (1c) = *Fred went to the supermarket because his fridge is empty. Then, he went to the movies.*

Next, discourse connectives in D-STAG all anchor elementary trees with **two** arguments, whereas, in D-LTAG, most adverbial connectives (but not the empty one ϵ) anchor trees with only **one** argument (this is, for example, the case for *then* whose host segment is "structurally" provided, the mate segment being "anaphorically" provided [23]). Moreover, subordinating conjunctions in D-LTAG anchor trees with two arguments, but these trees are **initial** trees whereas they are **auxiliary** in D-STAG (subordinating conjunctions are considered in D-LTAG as "structural" connectives, neglecting the fact that one of their segments can cross a sentence boundary, (1a) and (3)). These discrepancies between the trees anchored by a connective lead to crucial differences in the discursive analyses, especially in the semantic ones.

As an illustration, the syntactic tree and derivation tree produced by D-LTAG for (1c) are given in Fig. 13. The syntactic tree includes three nodes labelled DC, while it includes only two nodes labelled DC in D-STAG since the empty connective ϵ is introduced only in the absence of any other adverbial. The derivation tree results in the discourse structure $Narration(Explanation(F_0, F_1), F_2)$, which is wrong: the explanation given for Fred's visit to the supermarket (*his fridge was empty*) shouldn't be under the scope of *Narration*, which is a relation linking together two events and not a causal relation and an event. As explained in Sect.2, the discourse structure for (1c) is $Explanation(F_0, F_1) \land Narration(F_0, F_2)$. This structure, which corresponds to a non tree-shaped dependency graph, cannot be obtained in D-LTAG.

"Multiple connectives" is a phenomenon which guided the conception of D-LTAG, as shown in the quotation [23, p 552]: "The distinction between structural and anaphoric connectives in D-LTAG is based on considerations of computational economy and behavioral evidence from cases of multiple connectives". The discourse (6) gives an illustration of multiple connectives. Its DNF is of the form C_0 *but* C_1 *because then* C_2, in which two connectives share the same host clause C_2, the second one being an adverbial.

(6) John ordered three cases of Barolo. But he had to cancel the order *because then* he discovered he was broke. [23]

In D-STAG, we propose the following solution to handle multiple connectives. The DNF for (6) is automatically converted into C_0 *but* C_1 *because* $\overline{C_2}$ *then* C_2 which conforms to the regular pattern for DNFs (without any preposed conjunction).

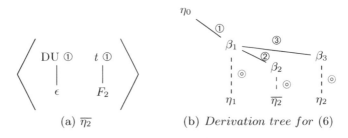

(a) $\overline{\eta_2}$ (b) *Derivation tree for* (6)

Fig. 14. Tree pair $\overline{\eta_2}$ and derivation tree for (6)

The tree pair $\overline{\eta_2}$ associated to $\overline{C_2}$ is given in Fig. 14-a: the syntactic leaf is the empty string ϵ, the semantic leaf is F_2 (the semantic formula for C_2). So we can say that $\overline{C_2}$ is the (fake) host clause of *because*, which has only a semantic contribution. The interpretation of (6), i.e. $Contrast(F_0, F_1) \wedge Explanation(F_1, F_2) \wedge Narration(F_0, F_2)$ assuming that *then* conveys *Narration*, is obtained by the mechanisms described in Sect. 5.1.1 which build the derivation tree given in Fig. 14-b (with $\beta_1 = but \div Contrast$, $\beta_2 = because \div Explanation$ and $\beta_3 = then \div Narration$).[9]

In conclusion, cases of multiple connectives can be boiled down to the general pattern in which an host clause receives only a single connective, thanks to the sentence-discouse interface (which in any case cannot be bypassed (Sect.4)). Therefore, there is no need to postulate a distinction between anaphoric and structural connectives, the former with a single argument, the latter with two arguments.

7   Conclusion

We focused on an STAG discursive grammar for connectives/discourse relations. It aims at handling any type of connectives (adverbials, postposed and preposed subordinating conjunctions, coordinating conjunctions (including coordinating conjunctions in correlative constructions), any modifier of a connective, and finally cases of "multiple connectives". The syntactic discursive grammar for connectives shows that it is justified to postulate that any connective anchors an auxiliary tree with two arguments: a foot node DU* for the mate argument , a substitution node DU↓ for the host argument which is identical to or starts at the host clause of the connective. The semantic discursive grammar is made up from trees anchored by functors associated to discourse relations, which are considered as predicates lexicalized by discourse connectives. It allow us to compute discourse structures which are not compulsorily represented as tree-shaped dependency graphs.

---

[9] The interpretation of (6) looks as a counter-example to RFC since the discourse relation *Contrast* is coordinating. However, we explain in [8] how to handle this apparent counter-example.

The discursive component takes as input a DNF that is computed by a sentence-discourse interface. This interface is necessary if one doesn't want to make any change to the sentential analyzer. A DNF is a sequence of discourse words which follows a regular grammar. The regular grammar we have presented in this paper focuses on connectives but is not yet completed. In particular, it doesn't take into account sub-clauses (clausal complements, incident clauses, relative clauses) which may play a role at the discursive level. This is ongoing work [9] and we don't foresee any crucial difficulty for the resulting extended discursive grammar, thanks to STAG's richness of expressivity.

The implementation of D-STAG in a French discourse analyzer is work in progress [8]. The analyzer will produce a forest of derivation trees which represents the set of possible analyses. The extraction of the best analysis (or the n best analyses) will require to build probabilistic disambiguation models based on the French annotated corpus Annodis [19].

References

1. Asher, N.: Reference to Abstract Objects in Discourse. Kluwer, Dordrecht (1993)
2. Asher, N., Lascarides, A.: Logics of Conversation. Cambridge University Press, Cambridge (2003)
3. Asher, N., Vieu, L.: Subordinating and coordinating discourse relations. Lingua 115(4), 591–610 (2005)
4. Charolles, M.: Framing adverbials and their role in discourse cohesion, from connection to foward labelling. In: Proceedings of the First Symposium on the Exploration and Modelling of Meaning, SEM 2005, Biarritz, pp. 194–201 (2005)
5. Danlos, L.: A lexicalized formalism for text generation inspired from TAG. In: Abeillé, A., Rambow, O. (eds.) TAG Grammar. CSLI, Stanford (2001)
6. Danlos, L.: Discourse dependency structures as constrained DAGs. In: Proceedings of SIGDIAL 2004, Boston, pp. 127–135 (2004)
7. Danlos, L.: Strong generative capacity of RST, SDRT and discourse dependency DAGs. In: Benz, A., Kühnlein, P. (eds.) Constraints in Discourse, Benjamins (2007)
8. Danlos, L.: D-STAG: un formalisme d'analyse automatique de discours basé sur les TAG synchrones. Revue TAL 50(1), 111–143 (2009)
9. Danlos, L., Sagot, B., Stern, R.: Analyse discursive des incises de citations. In: Actes du Second Colloque Mondial de Linguistique Française New-Orleans, USA (submitted)
10. Forbes-Riley, K., Webber, B., Joshi, A.: Computing discourse semantics: The predicate-argument semantics of discourse connectives in D-LTAG. Journal of Semantics 23(1) (2006)
11. Harris, Z.: Mathematical Structures of Language. Krieger Pub. Co., New York (1986)
12. Joshi, A.: Tree-adjoining grammars. In: Dowty, D., Karttunen, L., Zwicky, A. (eds.) Natural Language Parsing, pp. 206–250. Cambridge University Press, Cambridge (1985)
13. Lee, A., Prasad, R., Joshi, A., Webber, B.: Departures from tree structures in discourse: Shared arguments in the penn discourse tree bank. In: Proceedings of the Constraints in Discourse Workshop (CID 2008). Postdam, Germany (2008)

14. Mann, W.C., Thompson, S.A.: Rhetorical structure theory: Toward a functional theory of text organization. Text 8(3), 243–281 (1988)
15. Marcu, D.: The rhetorical parsing of unrestricted texts: A surface-based approach. Computational Linguistics 26(3), 395–448 (2000)
16. Marcu, D.: The Theory and Practice of Discourse Parsing and Summarization. The MIT Press, Cambridge (2000)
17. Nesson, R., Shieber, S.: Simpler TAG semantics through synchronization. In: Formal Grammars, Malaga (2006)
18. PDTB Group: The penn discourse treebank 2.0 annotation manual. Tech. Rep. IRCS-08-01, Institute for Research in Cognitive Science, University of Philadelphia (2008)
19. Péry-Woodley, M.P., Asher, N., Enjalbert, P.: ANNODIS: une approche outillée de l'annotation de structures discursives. In: Proceedings of TALN 2009, pp. 190–196. Senlis, France (2009)
20. Schabes, Y., Waters, R.: Tree insertion grammar. Computational Intelligence 21, 479–514 (1995)
21. Shieber, S.: Restricting the weak-generative capacity of synchronous tree-adjoining grammars. Computational Intelligence 10(4), 371–385 (1994)
22. Shieber, S., Schabes, Y.: Synchronous tree-adjoining grammars. In: Proceedings of the 13th International Conference on Computational Linguistics, vol. 3, pp. 253–258. Helsinki, Finland (1990)
23. Webber, B.L., Joshi, A., Stone, M., Knott, A.: Anaphora and discourse structure. Computational Linguistics 29(4), 545–587 (2003)
24. Wolf, F., Gibson, E.: Coherence in Natural Language: Data Structures and Applications. The MIT Press, London (2006)

An Efficient Enumeration Algorithm for Canonical Form Underspecified Semantic Representations

Mehdi Manshadi, James Allen, and Mary Swift

University of Rochester,
Rochester, NY
{mehdih,james,swift}@cs.rochester.edu

Abstract. We give polynomial-time algorithms for satisfiability and enumeration of underspecified semantic representations in a *canonical form*. This canonical form brings several underspecification formalisms together into a uniform framework (Manshadi et al., 2008), so the algorithms can be applied to any underspecified representation that can be converted to this form. In particular, our algorithm can be applied to Canonical Form Minimal Recursion Semantics (CF-MRS). An efficient satisfiability and enumeration algorithm has been found for a subset of MRS (Niehren and Thater, 2003). This subset, however, is not broad enough to cover all the meaningful MRS structures occurring in practice. CF-MRS, on the other hand, provably covers all MRS structures generated by the MRS semantic composition process.

Keywords: Formal Semantics, Underspecification, Minimal Recursion Semantics, Canonical Form Underspecified Representation.

1 Introduction

Underspecification in semantic representation is about encoding semantic ambiguities in a semantic representation. Efficient enumeration of all possible readings of an underspecified semantic representation is a topic that has interested researchers since the introduction of underspecification in semantic representation. Hobbs and Shieber (1987) is one of the earliest works on this topic. The underspecification formalism that they use is based on a traditional underspecified logical form (Woods 1978), which is neither flat nor constraint-based. Most of the recent semantic formalisms, however, use a flat, constraint-based representation of natural language semantics, such as *Minimal Recursion Semantics* (Copestake et al., 2001), *Hole Semantics* (Bos 1996), and *Dominance Constraints* (Egg et al., 2001).

Recently there has been some work on finding efficient algorithms for determining whether an underspecified representation has a reading or not (the *satisfiability* problem) and for enumerating all the possible readings of a satisfiable representation (the *enumeration* problem). Althaus et al. (2003) shows that the satisfiability problem for Dominance Constraints formalism in its general form is NP-complete. Niehren and Thater (2003) define a subset of dominance constraints called *dominance nets*, and show that an algorithm given by Bodirsky et al. (2004) can be used to generate the readings of a dominance net. Furthermore, they define a translation of *Minimal*

P. de Groote, M. Egg, and L. Kallmeyer (Eds.): FG 2009, LNAI 5591, pp. 85–101, 2011.
© Springer-Verlag Berlin Heidelberg 2011

Recursion Semantics(MRS) to dominance constraints. As an analogy to dominance nets, they also define a subset of MRS called *MRS nets* and prove that there is a bijection between the readings of a MRS net and the readings of its corresponding dominance net. This shows that the above mentioned algorithm can be used for enumeration of MRS nets, a big subset of MRS. They do not, however, make any claim about the coverage of MRS nets. By studying the output of *the LinGOEnglish Resource Grammar (ERG)* (Copestake and Flickinger 2000) on the *Redwoods* Treebank (Oepen et al., 2002), Fuchss et al. (2004) claim that all the non-net MRS structures are semantically "incomplete". In other words, they claim that the concept of net is broad enough to cover all semantically complete MRS structures. This claim, however, later was invalidated (Thater 2007). That is there are examples of coherent English sentences whose MRS structure is not a net (see section 6). As a result no efficient enumeration algorithm has been found that covers all MRS structures occurring in practice.

In recent work, Manshadi et al. (2008) define another subset of MRS structures called *Canonical Form* MRS *(CF-MRS)* and prove that it covers all the well-formed MRS structures generated by the *MRS semantic composition algorithm* (Copestake et al. 2005). Motivated by the definition of CF-MRS, they define a *Canonical Form Underspecified Representation (CF-UR)* and claim that this representation can be translated back and forth to some other underspecification formalisms, such as Dominance Constraints and Hole Semantics.

In this paper, we give a polynomial-time algorithm for satisfiability and enumeration of CF-UR. This directly results in an efficient algorithm for solving CF-MRS. Since CR-MRS has been proved to cover every well-formed MRS generated by the MRS semantic composition process, our algorithm covers coherent non-net examples that previously proposed algorithms do not.

The structure of this paper is as follows. We give an informal introduction in to CF-UR in (2.1) and the formal definition in (2.2). We define *dependency graph* (3) and present the algorithms to solve a *first-order* CF-UR (4) and the CF-UR in its general form (5). (6) discusses the related work in detail; it specifically addresses the difference between MRS net and CF-MRS.

2 Canonical Form Underspecified Representation

2.1 An Informal Introduction

We first explain the concept of CF-UR through an example. Consider the sentence *Every dog probably chases some cat*. Two of its readings are shown in figure (1), and the corresponding CF-UR in graphical form is shown in figure (2).

(a)   (b)

Fig. 1. *Two readings of the sentence* Every dog probably chases some cat

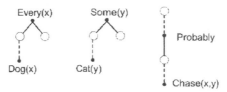

Fig. 2. CF-UR graph

As shown in this figure, the graph has two types of node: *label node* and *hole node,* and two types of edge: *solid edge* and *dotted edge.* The graph is directed, although the directions are not shown explicitly when the graph is a tree or a forest. There are three kind of label nodes: first-order predicates (such as *Dog(x), Chase(x,y)*), operators (such as *Probably*) and quantifiers (such as *Every(x)*). Every quantifier node has two outgoing solid edges to two hole nodes: one for its restriction (*restriction hole*) and one for its body (*body hole*). Operators such as *Probably* also have one or more outgoing solid edges to distinct hole nodes. There is only one hole node in the graph which does not have any incoming edge. This is called the *top hole.* As seen in figure (2), the number of hole nodes and label nodes in the graph are equal. In order to build the readings of a CF-UR, we must plug the label nodes into the hole nodes. But not every plugging[1] is desirable. The dotted edges in a CF-UR represent the *qeq* (equality modulo quantifiers, Copestake et al., 2005) *constraints.* A qeq constraint from hole *h* to label *l* is satisfied, if either the label *l* directly plugs into the hole *h* or, *h* is filled by a quantifier node and *l* plugs into the body hole of that quantifier, or there is another quantifier which plugs into the body hole of the first one and *l* plugs into the body of the second one and so on. Given a CF-UR, a tree is built by removing all the constraint edges and plugging every label to a distinct hole. The tree is called a fully scoped structure iff it satisfies all the qeq constraints. For example, figures (1a,b) both satisfy the four constraints of the CF-UR in figure (2); hence they are fully scoped structures of this CF-UR.

To have a valid reading of an underspecified representation every variable must be in the scope of its quantifier. We call this *dependency constraint.* A node *P(...)* is *dependent* on a quantifier node *Q(x)* if *x* is an argument of *P(...).* A fully scoped structure satisfies this constraint iff *Q(x)* outscopes[2] (i.e. is an ancestor of) *P(...)* in the tree. A fully scoped structure of a CF-UR is called a *reading* or a *solution* iff it satisfies all the dependency constraints. Manshadi et al. (2008) prove that every well-formed MRS structure generated by the MRS's semantic composition process is in a canonical form (hence called CF-MRS), which is a notational variant of CF-UR. Furthermore, they show that in a CF-MRS, qeq relationships and outscoping constraints are equivalent; that is if the qeq constraints in a CF-UR are treated as outscoping constraints, the CF-UR will still have the same set of readings. As mentioned before, the dependency constraints are also outscoping relations; therefore similar to the Dominance Constraints formalism, the dependency constraints are made explicit (figure 3), and all the constraints in CF-UR are treated as outscoping relations.

---

[1] We borrowed the term plugging from Hole Semantics. Manshadi et al. (2008) call the plugging a label assignment.

[2] Note that outscoping (or dominance) is considered to be a reflexive and transitive relation.

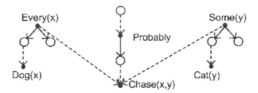

Fig. 3. UR graph with explicit dependencies

2.2 The Formal Definition

Consider *F* the set of labeled formulas of the following types:

- *Quantification:* A formula of the form *l:Q(x, hr, hb)* where *l* is the label, *Q* is the generalized quantifier, *x* is the first order variable quantified by *Q*, and *hr* and *hb* are the holes for the *restriction* and the *body* of the quantifier.

- *Operators:* A formula of the form *l:P(x1, x2, ..., h1, h2, ...)* where *P* is an operator, *x1, x2, ...* are first order arguments of *P* and *h1, h2, ...* are holes for the higher order arguments.

- *Predications:* A formula of the form *l:P(x1, x2, ...)* where *P* is a first order predicate and *x1, x2, ...* are its arguments.

A CF-UR is the triple $U = <F, h_T, C>$ in which *F* is a set of labeled formulas, h_T is a unique hole which does not occur in any argument position in *F*, called the *top hole*, and $C=C_q \cup C_d$, where C_q is a set of hole to label constraints (corresponding to qeq relationships) and C_d is a set of label to label constraints (corresponding to dependency constraints). We require *U* to satisfy following conditions (the canonical form conditions):

a) No quantifier labels and no quantifier body holes are involved in any constraint in C_q.

b) Every other hole and label is involved in exactly one constraint in C_q.

A *label assignment* or *plugging P* is a *bijection* between holes and labels. The ordered pair *<U, P>* is called a *reading* or a *solution* of *U* iff it satisfies all the constraints in $C=C_q \cup C_d$.

<U, P> satisfies an outscoping constraint $u \le v$ iff

- when *u* and *v* are both labels: *u=P(..., h, ...)* is in *F* and $h \le v$ recursively holds.

- when *u* is a hole and *v* is a label: *P(u)=v* or *P(u)=l*, where *l=P(..., h, ...)* is a labeled formula in *F* and $h \le v$ recursively holds.

As an example, the CF-UR for the sentence *Every dog probably chases some cat* is shown below.

U = <{l1: Every(x, h1, h2), l2: Dog(x), l3: Some(y, h3, h4), l4: Cat(y), l5:Probably(h5), l6:Chase(x,y)}, **h0**, *{h0≤l5, h1≤l2, h3≤l4, h5≤l6} ∪{l6≤l1, l6≤l3}>*

Figure (3) shows a graphical representation of the CF-UR *U*, in which the dotted edges represent the constraints (the labels of the formulas are not shown in this figure). Note that the hole nodes of every formula are assumed to be ordered from left to right. A plugging *P*, which satisfies all the constraints in *U*, is given below.

$$P = \{(h0, 13), (h1, 12), (h2, 16), (h3, 14), (h4, 15), (h5, 11)\}$$

P corresponds to the graphical representation of the solution $<U,P>$ in figure (1b) above, which is obtained by removing all the dotted edges from U's graph in figure (2) and merging every hole node h with the label node $P(h)$.

A CF-UR is called *satisfiable* iff it has at least one solution. The problem of finding all possible solutions of a CF-UR is called the *enumeration* problem.

Since dependency constraints are obvious from the first order arguments of the formulas, for the sake of readability, we often remove the dependency constraints from CF-UR's graph as in figure (2) above. The graph in figure (2) is a forest of three trees; two of which are rooted at the two quantifiers and one rooted at the top hole. Using canonical form conditions (conditions (a) and (b) given above), it is easy to see that this configuration holds in general; that is given a CF-UR U with n quantifier nodes, if we ignore the dependency constraints (i.e. C_d), U's graph is a forest of exactly $n+1$ trees whose roots are the top hole and the quantifiers as in figure (4).

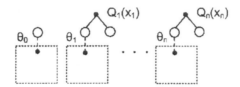

Fig. 4. General structure of CF-UR

3   Dependency Graph

In this section we build a mathematical framework to formally present the algorithms and prove their properties. We will first solve the satisfiability and enumeration problems for a CF-UR with only quantifiers and first order predicates (i.e. no operators). We call such a CF-UR a *first-order* CF-UR (figure 5).

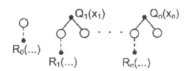

Fig. 5. General structure of first-order CF-UR

To generalize this to handle an arbitrary CF-UR U with operators (figure 4), we transform U into a first-order CF-UR U', which we call U's *reduced form*, by collapsing the trees θ_i $(i=0..n)$ into a single first order predicate $R_i(...)$, whose arguments are the union of the first order arguments of all the predicates and operators in the tree θ_i (figure 6). We use the algorithms for first-order CF-UR (given in section 4) to solve U' and use the results to solve the original problem (section 5).

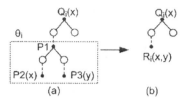

Fig. 6. Collapsing the trees into a single node

Consider a first order CF-UR with n quantifiers (figure 5). In order to build all its corresponding fully scoped structures, all we need to know is the number of quantifiers. For example, a first order CF-UR with two quantifiers has four possible fully scoped structures shown in figure (7).

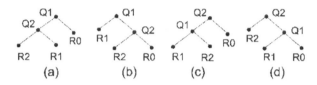

Fig. 7. All fully scoped structures for n =2

In order to check whether each fully scoped structure is a solution or not, we only need to check the satisfaction of the dependency constraints. We define the concept of *dependency graph* to represent all these dependencies in a compact form. A *Dependency Graph (DG)* is a directed graph with $n+1$ nodes labeled $0...n$, corresponding to the nodes R_0 to R_n. The node i $(i>0)$ is connected to node j by a directed edge (i, j) iff R_j is dependent on Q_i. In all the examples in this paper, we assume that the quantifiers are numbered in the order they appear in the sentence. As an example consider the CF-UR in figure (8) for the sentence *Every politician whom somebody knows a child of runs*. The DG for this sentence is given in figure (9c).

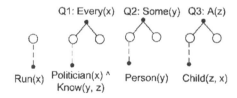

Fig. 8. *CF-UR for* Every politician whom somebody knows a child of runs

Figures (9a,b) show the DGs for the sentences *Every dog chases a cat* and *Every dog in a room barks* respectively. Note that out of the four possible fully scoped structures for n=2, given in figure (7), (7b,d) are the solutions for the DG in (9a) and (7a,d) are the solutions for DG in (9b).

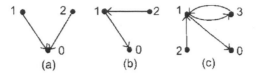

Fig. 9. Examples of dependency graph

From the definition, node 0 in every DG has no outgoing edges, so it is a *sink* node. We call the sink node the *heart* of the DG. In all the DGs in figure (9), every node can reach the heart by a directed path. We call this property (that every node in a DG is connected to the heart by a directed path) *heart-connectedness*. Therefore all three DGs in figure (9) are *heart-connected*. This is not a coincidence. Note that if a node is directly connected to the heart by a directed edge, it means that the corresponding noun phrase (NP) is an argument of the heart formula (i.e. the main predicate of the sentence). If a node is connected to the heart by a path of length two, it means that the corresponding NP is a modifier of an argument of the heart formula, and so on. The heart-connectedness property requires that every NP contribute to the meaning of the sentence, either by filling an argument position of the main predicate or by modifying an argument of the main predicate, and so on; which is a trivial property of every coherent sentence.

4   Scoping in First Order CF-UR

Consider a first order CF-UR U with the heart-connected DG G. A solution of U (or simply a solution of G) is an ordered binary tree T with exactly n interior nodes labeled $Q_1..Q_n$ and $n+1$ leaf nodes labeled $R_0...R_n$ such that

- *Qeq constraints:* R_0 is the right-most leaf of T. Every other R_i is the right-most leaf of the tree rooted at the left child of Q_i in T.
- *Dependency constraints:* for every i, j $(i>0)$, if node i *immediately dominates* node j in G, then Q_i dominates R_j in T.

Here, by u *immediately dominates* v, we mean u is connected to v by an edge (u,v). *Dominates* is the *reflexive transitive* closure of *immediately dominates*. We represent the tree rooted at the left child of Q_i in T, T_i and call it the *restriction tree* of Q_i. Similarly, the tree rooted at the right child of Q_i is called the *body tree* of Q_i

For example, the DG given in figure (10a) for the sentence *Every dog in a room of some house barks* has 5 different solutions, two of them given in figure (10 b,c).

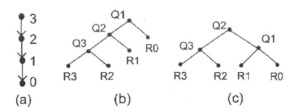

Fig. 10. A DG with two of its solutions

It is important to understand some counter intuitive properties of DG. First, the fact that i immediately dominates j in G, although implying Q_i has to dominate R_j in T, does not necessarily mean that Q_i has to dominate Q_j in T. This is shown in both figures (10b,c) for the nodes $Q2$ and $Q3$. Second, if i *transitively* (i.e. not immediately or reflexively) dominates j in G, it does not force Q_i to dominate R_j in T, as shown for nodes $Q3$ and $R1$ in both figures (10 b,c).

4.1 Satisfiability Algorithm

If the DG G has no *directed cycle* (hence is a *directed acyclic graph* or *DAG*), then there is a *topological order* of the nodes in G, say $n, n-1, ..., 0$. In this case, figure (11) is an interpretation of G. Hence G is trivially satisfiable.

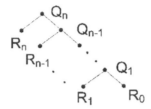

Fig. 11. An interpretation for a DAG

Now consider the DG G in figure (9c) as an example of a DG with directed cycles. We can partition G into 3 *Strongly Connected Components* (*SCC*) G_0, G_1, G_2 as shown in figure (12a). A node in an SCC, which connects it to some node outside the SCC, is called a *head* of SCC. In this example, every SCC (except G_0) has exactly one head. We replace every SCC (except G_0) with its head as shown in figure (12b) and call the new graph G'. G' is a DAG, so we can build an interpretation T' of G' as shown in figure (12c). Since we replaced G_1 with a single node, Q_3 is missing in T'. We remove all the outgoing edges of the head (node 1) in G_1 and call the new graph G'_1 as shown in figure (12d). If we treat node 1 as the heart, G'_1 can be seen as a heart-connected DG. Therefore we can recursively apply the same procedure to G'_1 to build a solution T_1 shown in figure (12e). When we return from this recursive procedure, we replace the node R_1 in T' with the tree T_1 and call the new tree T (figure 12 f). T is a solution of G.

Note that if G_1 has more than one head, that is if node 3 is also directly connected to the heart, then T would no longer be a solution of G, as R_0 is not in the scope of Q_3 in T. This suggests rejecting every DG containing an SCC with more than one head.

To state this idea formally, consider a DG $G=(V, E)$, with $K+1$ SCC $G_0...G_K$ where G_0 only contains the heart node. We formally define a head of $G_k=(V_k, E_k)$ as a node u in V_k such that there exists a node v in $V-V_k$ where (u,v) is an edge in E. Since G is heart-connected, every G_k $(k>0)$ has at least one head. We call a DG G *single-headed* iff every SCC G_k has at most one head. Let i_k $(k>0)$ be the head of G_k. We remove all the outgoing edges of i_k in G_k to call the resulting graph G'_k. If we treat i_k as the heart, G'_k can be seen as an independent DG, called the *underlying DG* of G_k.

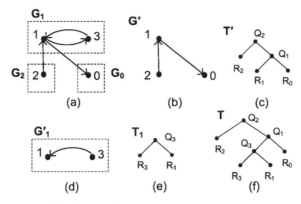

Fig. 12. Satisfiability of a DG with cycles

SAT(G)
1. *Find all the SCCs in G ($G_0...G_K$).*
2. *If some SCC has more than one head output UNSAT and halt, otherwise:*
3. *Replace each SCC G_k (k>0) with its head i_k and call the new graph G'.*
4. *Find a topological order of the nodes in G', say $i_K .. i_1$; build an interpretation T' of G' as shown in figure (14a).*
5. *For each SCC G_k with more than one node, remove all the outgoing edges of the head and call it G'_k. Let $T_k = SAT(G'_k)$*
6. *Replace every Ri_k in T' with T_k; call the new tree T as shown in figure (14b).*
7. *Return T*

Fig. 13. SATisfiability algorithm for first-order CF-UR

A heart-connected DG G is called *recursively single-headed* iff

i) G is a single SCC, or
ii) G is single-headed and the *underlying* DG of all its SCCs are recursively single-headed.

Note that if a single-headed G is heart-connected, all the underlying DGs of G are also heart-connected; therefore the concept of recursive single-headedness is well-defined.

Theorem 1. A heart-connected DG is satisfiable if and only if it is recursively single-headed.

Note that theorem 1 has an intuitive linguistic explanation. Since an SCC in the DG of a sentence represents a noun phrase and the head of an SCC actually represents the head of the noun phrase, this theorem says that an underspecified semantic representation has a reading if and only if every noun phrase has a single head.

The algorithm in figure (13) generalizes the procedure introduced in the above example. It returns a tree T if G is recursively single-headed and outputs UNSAT otherwise. To prove the *if* direction of theorem 1, all we need is to show that if the algorithm returns a tree T, T is a solution of G. We prove this using induction on the depth of the recursion d. For $d=0$, G is a DAG; hence T is trivially a solution of G. Consider a DG G for which the depth of recursion is d $(d>0)$ and let T be the output of the above algorithm. T is a solution of G since:

- *Qeq constraints:* $R0$ is trivially the right-most leaf of T. Also, for every $k>0$, using the induction assumption, T_k is a solution of G'_k with the heart i_k; therefore Ri_k is the right-most leaf of T_k. All other qeqconstraints hold by induction assumption.

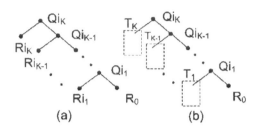

Fig. 14. Graphical description of the algorithm

- *Dependency constraints:* consider the nodes i, j such that i immediately dominates j in G. If i, j are in the same SCC, say G_k, then this dependency constraint holds by the induction assumption (because T_k is a solution of G'_k). If i, j are in two different SCCs, say G_k and G_l respectively, then we have $i=i_k$ (because i_k is the single head of G_k) and $i_k<i_l$ (because i_K, i_{K-1}, ... is a topological order); therefore Qi_k dominates Qi_l and the whole tree T_l *in T* (figure 14b) and since R_j is in T_l , Q_i dominates R_j in T.

The proof of the *only if* direction is not as straight forward. The proof idea is given below by stating three helpful lemmas. For a given node v in G, we define $Anc(v)$ as the set of nodes that dominate v (including v itself) and $Dis(v)$ as the set of nodes that have a path to the heart without going through v (including the heart itself). As G is heart-connected, for every v in V, we have $Anc(v) \cup Dis(v) = V$.

Lemma 1: If Q_i is the root of some arbitrary solution T of G, then for every $j \neq i$ in $Anc(i)$, Q_j is in the restriction tree of Q_i and for every $k>0$ in $Dis(i)$, Q_k is in the body tree of Q_i.

Proof: Consider $j \in Anc(i)$. We use induction on the length of the path P (shown as $/P/$) from j to i to show that Q_j is in the restriction of Q_i. First, let $/P/=1$, that is j immediately dominates i. Q_i is the root and R_i (the restriction of Q_i) has to be in the scope of Q_j, therefore Q_j has to be in the restriction tree of Q_i. Now Let $/P/=n+1$ $(n \geq 1)$ and k be the node immediately after j on the path P. According to the induction assumption, Q_k is in the restriction tree of Q_i. On the other hand, R_k must be in the scope of Q_j therefore Q_j has to be in the restriction tree of Q_i too. A similar argument applies when $j \in Dis(i)$. □

From this lemma, a node Q_i can be the root of some solution G only if $Anc(i)$ and $Dis(i)$ are disjoint. We call this property the *root condition*.

Lemma 2. If a subgraph G' of G (with the same heart) is unsatisfiable, then G is unsatisfiable.

Although simple, the above lemma is very helpful as it allows us to consider only the problematic part of the DG and ignore the rest. Now consider a DG G with an SCC G_1 which has at least two heads, say i_1 and j_1 (figure 15). The nodes i_1 and j_1 are connected to some node(s) outside SCC, say i and j respectively. As G is heart-connected, i and j are connected to the heart by paths P_1 and P_2. First, consider the case where the two paths intersect only at the heart node. Consider only the part of G which includes the SCC G_1 and the paths P_1 and P_2; call it G' (figure 15).

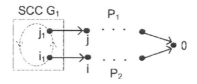

Fig. 15. An unsatisfiable DG

G' is unsatisfiable because none of the nodes in this graph satisfy the root condition. Therefore from lemma (2), G is unsatisfiable. A similar argument can be given for the case where P_1 and P_2 intersect at some other nodes as well. This completes the proof that G is satisfiable only if it is single-headed. To prove that G needs to be recursively single-headed, we use following lemma, which directly results from lemma 1.

Lemma 3. If i_k is the head of some SCC G_k and T is an arbitrary solution of G, for every node $j{\neq}i_k$ in G_k, Q_j is in the restriction tree of Qi_k in T.

From this lemma, the nodes in every SCC G_k form a smaller satisfiability problem whose DG *contains* the underlying DG of G_k (i.e. G'_k) and whose heart is the same as G'_k's heart (i.e. Ri_k). Therefore from lemma 2 the property of being single-headed must recursively hold for G'_k.

The algorithm given in figure (14) divides the satisfiability of G into K subproblems, satisfiability of $G'_1...G'_K$, where $|V'_1|+...+|V'_K|<|V|$ ($|G|$ is the size of G and $|V|$ is the number of nodes in G). The cost of this breaking is linear in $|G|$. Therefore if $T(|G|)$ is the running time of the algorithm, we have:

$$T(|G|) = O(|G|)+T(|G'_1|)+...+T(|G'_K|)$$

Using induction on $|V|$, the worst-case complexity of the algorithm is quadratic in size of G. More precisely the running time is $O(|V|{\cdot}|G|)$, where $|G| = |V|+|E|$ if we represent the graph using adjacency list.

4.2 Enumeration Algorithm

Using lemma 1, we showed that Q_i can be a root of some solution T of G only if $Anc(i)$ and $Dis(i)$ are disjoint. Consider the subgraph of G induced by the nodes in $Dis(i)$ and call it G_b (figure 16 is an example where $i{=}1$).

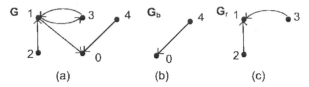

Fig. 16. G, G_r, and G_b *for* Every politician whom somebody had a chat with voted for the bill

From lemma 1, it can be seen that the body tree of Q_i in T has to be a solution of G_b. Now, consider the subgraph of G induced by the nodes in $Anc(i)$. We remove all the outgoing edges of i and call the resulting DG, G_r (figure 16c). From lemma 1, the restriction tree of Q_i in T has to be a solution of G_r. As a result, Q_i is a root of some solution of G if and only if it satisfies the root condition and its corresponding G_r and G_b are both satisfiable. Note that since G is heart-connected, both G_r and G_b are heart-connected. Figure (17) summarizes the enumeration algorithm.

EnumerateFO (G)
1. *If G is not satisfiable fail.*
2. *If G has only one node, return G.*
3. *Find R, the set of nodes i in G which satisfy the root condition.*
4. *Non-deterministically pick a node i in R:*
5. *Build G_r and G_b*
6. *Let T_r = EnumerateFO(G_r) & T_b = EnumerateFO(G_b)*
7. *Build the tree T rooted at Q_i, with T_r and T_b as restriction and body tree of Q_i respectively.*
8. *Return T*

Fig. 17. Enumeration for first-order CF-UR

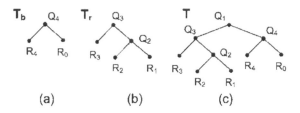

Fig. 18. An example of enumeration algorithm

As an example, if the recursive call to *EnumerateFO()* for G_r and G_b in figure (16), returns the trees T_r and T_b in figure (18a,b), the final solution tree T would be the one in figure (18c). An argument similar to the proof of the *if* direction of theorem 1 can be given to prove the soundness of the algoithm. The completeness can be prove by induction. If T is a solution of a heart-connected DG G rooted at Q, according to the above discssions and lemma 1, Q must satisfy the root condition (hence it will be picked as the root at some branch of the algorithm). On the other hand, the restriction and the body tree of Q must be solutions of G_r and G_b (hence they are built at step 5 based on induction asumption), therefore T will be generated by the algorithm.

The enumeration algorithm breaks the problem into subproblems, but this time the cost of this breaking is quadratic in $/G/$ (because we check the satisfiablity at each step). Therefore the time complexity of the overall procedure is cubic in size of G per *solution*. The time-complexity can be improved though. It can be shown that the satisfiability check at each step is not necessary. In fact, we can remove step 1 of the algorithm and if the enumeration algorithm ever fails (i.e. it encounters an empty R) we declare G as unsatisfiable.After this simplification, the running time would be $O(/V/ \cdot /G/)$. Space does not permit us to give the technical details of the proofs given here. We reserve those for a longer paper.

5 Scoping with Operators

The following theorem shows that to check the satisfiability of a CF-UR, it is enough to check the satisfiability of its reduced form.

Theorem 2. A CF-UR is satisfiable if and only if its reduced form is satisfiable.

The *only if* direction is trivial. The *if* direction is true because for every solution of the reduced form, at least one solution of the original CF-UR can be built by taking the solution T of the reduced form, expanding every node R_i to its original tree θ_i (see figures 4-6) and simply assigning every label in the θ_i to the hole to which the label is qeq. We call such solutions *basic* solutions. For example, figure (19a) shows the CF-UR for the sentence *Every dog which probably chases some cat does not bark* with its reduced form in figure (19b). Figures (19 c,d,e) show one solution of the reduced form, its expanded version, and the corresponding basic solution of the original MRS respectively. In general, corresponding to each solution of the reduced form, there is more than one solution of the original CF-UR. For example there are three more solutions corresponding to (19c) as shown in (19 f,g,h).

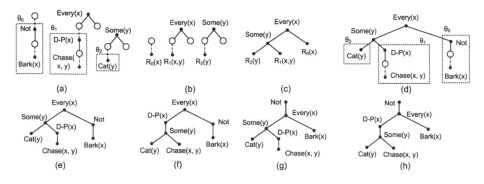

Fig. 19. Scoping with operators

The solutions other than basic ones are built by taking a basic solution and moving quantifiers with their restriction trees inside operators (figure 20).

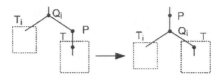

Fig. 20. Moving quantifiers inside operators

A quantifier Q_i can move inside an operator P only if P is not dependent on Q_i. The enumeration algorithm for a general CF-UR is shown in figure (21). For every basic solution T', *EnumerateB(T', n)* is called to build all the corresponding non-basic solutions. In order to prevent generating a single solution in more than one way, quantifiers in a basic solution are ordered in a post order fashion and are picked by *EnumerateB()* based on this order using argument m (see condition (a) in step 2 of *EnumerateB()*). Trivially the algorithm is sound. To see why it is complete, consider an arbitrary solution T (e.g. figure 19h); move quantifiers with their restriction trees (based on a preorder) all the way up in T until it hits another quantifier node; the resulting tree T' would be a basic solution (figure 19e).

Enumerate(U)   *// U: a CF-UR with n quantifiers*
   1.   Let G = DG of U' (Reduced-form of U)
   2.   Let T = EnumerateFO(G)
        T' = Basic solution corresponding to T
   3.   Call EnumerateB(T', n)

EnumerateB(T, m)
   1.   Output T
   2.   Non-deterministically pick a quantifier Q_i;
        a) whose order k is at most m;
        b) whose body node is an operator P;
        c) where P is not dependent on Q_i
   3.   Move Q_i inside P and call the new tree T'.
   4.   Call EnumerateB(T', k)

Fig. 21. General enumeration algorithm

Every branch of *EnumerateB()* takes linear time and *uniquely* generates a solution; therefore *EnumerateB()* runs in linear time per solution; therefore *Enumerate()* runs in quadratic time per solution as a result of the call to *EnumerateFO()*.

In these algorithms, we only considered single-hole operators. The extension to general case is straightforward; if an operator P has more than one hole, when moving a quantifier inside P, we non-deterministically pick a child of P and move the quantifier inside P along that child. We define an order on the children of P (e.g. from left to right) to prevent generating a single tree in exponentially many ways.

6 Related Work

There has been some work on satisfiability and enumeration of underspecified representation in the context of dominance constraints (Althaus 2003, Bodirsky 2004). Crucially, the concept of *solution* in that context has a different definition from the standard definition of reading in formal semantics, which we gave in section 2. In dominance constraints formalism, the standard notion of reading is referred to as the *constructive solution* or *configuration*. In the following discussion, we will use the term DC *solution* to refer to dominance constraints notion of solution and the term *reading* or *constructive solution* to refer to the standard notion of solution.

The main difference between a DC solution and a constructive solution is that there could be nodes in the DC solution that do not correspond to any label in the actual underspecified representation, but in a constructive solution every node corresponds to some label in the underspecified representation. As a result there are examples of underspecified representations that have DC solutions but no constructive solution. In fact, the problem of finding the DC solutions is easier than the problem of finding constructive solutions. However, even finding DC solutions for dominance constraints in general is NP-complete.

As a result, Althaus et al. (2003) define a subset of dominance constraints called *normal* dominance constraints and show that this subset can be solved in polynomial time. Bodirsky et al. (2004) expand the definition of normal dominance constraints to *weakly* normal dominance constraints and show that this larger subset still can be solved in polynomial-time. However, finding the constructive solutions of both normal and weakly normal dominance constraint is still NP-complete. This means that Bodirsky's algorithm cannot be used to find constructive solutions of weakly normal dominance constraints. Niehren and Thater (2003) define the concept of dominance net, a subset of weakly normal dominance constraints, and show that for this subset, Bodirsky's algorithm can be used to enumerate the constructive solutions.

As an analogy with dominance nets, they define a subset of MRS called MRS nets and show that Bodirsky's algorithm can be used to enumerate its readings. The concept of net, however, is too restricted. Figure (22) shows the three schemas that can occur in a net. In this figure dependency constraints are shown explicitly and all the constraint edges are interpreted as outscoping relations.

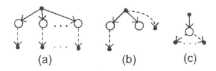

(a)   (b)   (c)

Fig. 22. Net schemas

The first schema corresponds to the operators in CF-UR where every hole has exactly one outgoing constraint. The second schema corresponds to the quantifier nodes where the restriction has one outgoing constraint edge and the body has no outgoing edge. The net condition requires that the quantifier node has exactly one outgoing dependency edge, which means there must be exactly one predicate (other than quantifier's restriction) dependent on every quantifier. This is where the limitation of nets comes from: not every natural language sentence satisfies this restriction. For example consider the CF-UR in figure (8). Figure (23) shows the same CF-UR with explicit dependency constraints.

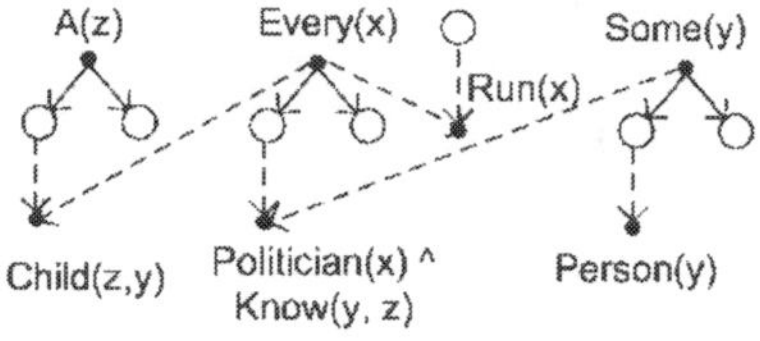

Fig. 23. A non-net CF-UR

As seen in this figure, the quantifier *Every* has two outgoing dependency edges, therefore it is not a net. Note that in CF-UR, there is no restriction on the number of outgoing dependency edges of a quantifier, therefore CF-UR covers the non-net structures such as the above example.

The schema in figure (22c), on the other hand, does not have any counterpart in CF-UR. This means that CF-UR lacks a certain kind of structure covered by nets. However, within the context of practical MRS structures this is not a limitation for CF-UR. In fact, this schema only occurs when translating MRS structures into dominance constraints. Since the dominance constraints formalism does not allow a free hole, the top hole of an MRS is replaced with a dummy operator *Prop* with a single hole. To enforce that this predicate has the widest scope (i.e. be the root of every reading), the hole is connected to every other label by a dominance edge (cf. Thater 2007). This results in one structure of schema 3 in every MRS net. No such a transformation is needed when MRS structures are represented in canonical form, therefore such a structure never occurs in CF-MRS. As a result, within the domain of practical MRS structures, nets are a strict subset of canonical form structures. After all, CF-MRS is proved to cover all well-formed MRS structures generated by the MRS semantic composition process.

7 Conclusion

We have presented algorithms for satisfiability and enumeration of CF-UR, an underspecified semantic representation in a canonical form. CF-UR is a notational variant of CF-MRS, which is the set of all well-formed MRS structures that can be generated by the MRS semantic composition algorithm. CF-MRS is broader than MRS nets, a previously defined subset of MRS for which satisfiability and enumeration algorithms have been found; broad enough to cover all the MRS structures occurring in practice.

In addition, CF-UR brings several different formalisms together into a uniform framework. Therefore, the proposed algorithms can be applied to any underspecified representation that can be transformed into CF-UR. For example, by using the concept of dependency graph, it is straightforward to show that the enumeration algorithm given here can replace the traditional wrapping algorithm (Woods 1987) to generate all the readings of a logical form.

The main drawback with both CF-UR and nets is that they do not allow holes to have more than one constraint edge, while some semantic constraints such as island constraints require additional outscoping constraints on the restriction hole of quantifiers. Presenting a version of the algorithms for this extended under specified representation remains future work.

References

Althaus, E., Duchier, D., Koller, A., Mehlhorn, K., Niehren, J., Thiel, S.: An efficient graph algorithm for dominance constraints. Journal of Algorithms 48, 194–219 (2003)

Bos, J.: Predicate logic unplugged. In: Proc. 10th Amsterdam Colloquium, pp. 133–143 (1996)

Bodirsky, M., Duchier, D., Niehren, J., Miele, S.: An efficient algorithm for weakly normal dominance constraints. In: ACM-SIAM Symposium on Discrete Algorithms. The ACM Press, New York (2004)

Copestake, A., Flickinger, D.: An open-source grammar development environment and broad-coverage English grammar using HPSG. In: Conference on Language Resources and Evaluations (2000)

Copestake, A., Lascarides, A., Flickinger, D.: An Algebra for Semantic Construction in Constraint-Based Grammars. In: ACL 2001, Toulouse, France (2001)

Copestake, A., Flickinger, D., Pollard, C., Sag, I.: Minimal Recursion Semantics: An Introduction. Research on Language and Computation 3(4), 281–332 (2005)

Egg, M., Koller, A., Niehren, J.: The constraint language for lambda structures. Journal of Logic, Language, and Information 10, 457–485 (2001)

Fuchss, R., Koller, A., Niehren, J., Thater, S.: Minimal Recursion Semantics as Dominance Constraints: Translation, Evaluation, and Analysis. In: ACL 2004, Barcelona, Spain (2004)

Hobbs, J., Shieber, S.M.: An Algorithm for Generating Quantifier Scopings. Computational Linguistics 13, 47–63 (1987)

Manshadi, M., Allen, J., Swift, M.: Toward a Universal Underspecifed Semantic Representation. In: 13th Conference on Formal Grammar (FG 2008), Hamburg, Germany (2008)

Niehren, J., Thater, S.: Bridging the Gap Between Underspecification Formalisms: Minimal Recursion Semantics as Dominance Constraints. In: ACL 2003, Sapporo, Japan (2003)

Oepen, S., Callahan, E., Flickinger, D., Manning, C.: LinGO Redwoods. A Rich and Dynamic Treebank for HPSG. In: Beyond PARSEVAL, LREC 2002, Las Palmas, Spain (2002)

Thater, S.: Minimal Recursion Semantics as Dominance Constraints: Graph-Theoretic Foundation and Application to Grammar Engineering. PhD Thesis, Universität des SaarlandesWoods (2007)

SaarlandesWoods, W.A.: Semantics and Quantification in Natural Language Questiom Answering. Advances in Computers 17, 1–87 (1978)

A Unified Account of Hausa Genitive Constructions

Berthold Crysmann

Universität Bonn & Universität des Saarlandes
Poppelsdorfer Allee 47, D–53115 Bonn
crysmann@ifk.uni-bonn.de

Abstract. In this paper I shall propose an analysis of the Hausa bound genitive marker which unifies its use in possessives and partitives with that in gerundive and pre-nominal adjectival constructions (see Newman, 2000 and Jaggar, 2001 for a detailed overview). I shall provide evidence that the bound genitive marker *-n/-r* is not simply an enclitic variant of the free form marker *na/ta* derived at a surface phonological level, but rather an affix attached in the morphology. Under the analysis proposed here, the free form is an instance of dependent marking, mainly used for possessive modifiers, whereas the bound form is an instance of head-marking, signalling the presence of an adjacent in-situ complement. The formal analysis, which is carried out in the framework of HPSG (Pollard and Sag, 1994), will make crucial use of type-raising in the sense of Kim and Sag (1995) and Iida et al. (1994), in order to model head-marking of possessives and pre-nominal adjectives on a par with complement-taking strong verbal nouns (gerunds). Furthermore, the head-marking approach to the bound linker also connects the presence vs. absence of this marker to a more general property of the language, namely direct object marking (Crysmann, 2005).

1 Major Functions of the Hausa Genitive Linker

Genitival constructions in Hausa[1] are used to express a wide range of syntactic functions: apart from possessives and partitives, genitive linkers are also employed to mark pre-nominal adjectival modifiers and certain gerunds, when followed by a direct object complement.

In principle, there are two distinct ways to effect genitive marking in this language: either by means of a free form marker *na* (masculine and plural) or *ta* (feminine) appearing on the dependent, or else by means of a suffixal (or enclitic) marker *-n/-r* attached to the head.

---

[1] Hausa is an Afroasiatic language spoken mainly in Northern Nigeria and bordering areas of Niger. Both tone (high, low, falling) and length (long vs. short) are lexically and grammatically distinctive. Following common practice, I shall mark low tone with a grave accent and falling tone with a circumflex. Vowels unmarked for tone are high. Length distinctions are signalled by double vowels.

P. de Groote, M. Egg, and L. Kallmeyer (Eds.): FG 2009, LNAI 5591, pp. 102–117, 2011.

1.1   Possessives

Possessives in Hausa can be formed either with a free or a bound linker. In both cases, the linker agrees with the head noun the possessive attaches to: while *ta* and *-r* are used with feminine singular head nouns only[2], *na* and *-n* are essentially default forms occurring with masculine and plural heads.

(1)   a. rìigaa ta Audù
         gown.f L.f Audu.m

      'Audu's gown'

      b. rìiga-r   Audù
         gown.f-L.f Audu.m

      'Audu's gown'

(2)   a. littaafii na   Kànde
         book.m L.m Kande.f

      'Kande's book

      b. littaafì-n   Kànde
         book.m-L.m Kande.f

      'Kande's book'

Use of the bound possessive linker imposes some strict adjacency requirements: if, e.g., a PP intervenes between the head noun and the possessive, use of the bound form is banned and the free form must be used instead.

(3)   a.   littaafì-n   Audù bisà   Sarkii
              book.m-L.m Audu about Emir

              'Audu's book about the Emir'

      b.   littaafì na   Audù bisà   Sarkii
              book.m L.m Audu about Emir

              'Audu's book about the Emir'

      c.   * littaafì-n   bisà   Sarkii Audù
              book.m-L.m about Emir Audu

      d.   littaafii bisà   Sarkii na   Audù
              book.m about Emir   L.m Audu

              'Audu's book about the Emir'

Possessives featuring the free marker can also be used independently:

(4)   Naa   karàntà na   Kànde
      1.sg.completive read   L.m Kànde

      'I read Kande's.'

---

[2] See below, though, for an additional phonological constraint applying to the bound marker *-r*.

In sum, Hausa recognises two distinct strategies for marking the possessive: a dependent-marking strategy where the genitive marker appears initially on the possessive modifier as a free form, and a head-marking strategy featuring a "genitive" suffix. While dependent-marking is essentially free in that the modifier can be either separated from the head or the head can be elided, bound affixal marking imposes a strict adjacency requirement on the dependent phrase. Independently of marking strategy, however, the post-nominal NP clearly has the status of a semantic modifier.

1.2   Objects of Strong Verbal Nouns

Another construction in which the genitive linker surfaces are nominal gerunds, a construction which, at first sight, does not seem to share too much similarity with the prototypical use of the genitive described above.

Hausa Tense Aspect-Mood (TAM) markers can be coarsely divided into two major classes: non-continuative TAM markers, including, inter alia, completive, future, and subjunctive on the one hand side, and the continuative on the other. Whereas non-continuative TAM markers uniformly select standard verbal complements, the continuative marker selects a special "gerundive" form. Furthermore, continuative markers are the only TAM category that can directly take dynamic nouns as complements, without having to resort to do-support.

Morphologically, we can identify essentially two major classes of gerunds in Hausa: *waa* (or weak) verbal nouns and non-*waa* (strong) verbal nouns (Newman, 2000). Weak (*waa*) verbal nouns represent the regular productive pattern for verbs in grades (paradigms) 1, 4, 5, 6, and 7 (Parsons, 1960). When no (direct) object follows, the gerund in these paradigms is productively formed by affixation of a suffix -`waa (with an initial floating low tone). If an object follows in situ, the form of *waa* verbal nouns is identical to that of a verb selected by non-continuative TAM markers, i.e. the gerundive suffix is dropped. Verbs in grade 2 and 3, however, do not have a weak (*waa*) verbal noun, so they use a strong form instead. Besides verbs from grades 2 and 3, some verbs from other grades have a secondary strong form that can be used instead of the regular weak *waa* verbal noun. Besides the absence of the suffix -*waa*, the most salient feature of the strong form is the presence of the genitive linker: if followed by a nominal or pronominal direct object, strong verbal nouns are obligatorily inflected with the genitive linker -*n/-r*. Object pronouns attached to strong verbal nouns are taken from the (low tone) genitive set, rather than the (polar) accusative set used with weak verbal nouns or normal verbs.

(5)   a.   Yaa   tàmbàyee -shì   / *-sà   .
           3.m.sg.completive ask   -him.acc   -him.gen

           'He asked him.'

      b.   Yaanàa   tàmbayà-r -sà   / *-shi.
           3.m.sg.continuative ask.f-L.f   -him.gen   -him.acc

           'He is/was asking him.'

Despite the difference in genitive marking, strong verbal nouns still bear the same semantic relationship to their direct object complements as the verbs they derive from, or even their weak verbal noun counterparts. This is most obvious with verbs having both strong and weak verbal noun form, such as grade 1 ɗinkàa.

(6) a.   Yaa   ɗinkà rìigaa.
           3.m.sg.completive sew   gown

           'He has sewn the gown.'

     b.   Yanàa   ɗinkà rìigaa.
           3.m.sg.continuative sew   gown

           'He is/was sewing the gown.'

     c.   Yanàa   ɗinkìn rìigaa.
           3.m.sg.continuative sew   gown

           'He is/was sewing the gown.'

If no direct object is present in situ, the genitive linker is illicit.

(7) a.   Yanàa   ɗinkìi.
           3.m.sg.continuative sew

           'He is/was sewing.'

     b.   Rìigaa cèe   yakèe   ɗinkìi.
           gown   FOCUS 3.m.sg.continuative sew

           'It's the gown he is/was sewing.'

Given that genitive head-marking on strong verbal nouns is triggered by what is unmistakably an argument of the verb, I will take this as initial evidence that genitive head-marking in general may be better understood in terms of head-complement rather than head-adjunct relations.

1.3 Pre-nominal Adjectives

Adjectival modifiers in Hausa can appear in one of two structural positions: either following the head noun, or immediately preceding it. Both pre- and post-nominal adjectives agree with the head noun in number and gender. According to Newman (2000), the absence of inherent gender and the prevalence of agreement gender constitute one of the main pieces of evidence to postulate a distinct category of adjectives in this language, the other piece of evidence being their inherently attributive nature. While post-nominal adjectives do not trigger any further special morphological marking, pre-nominal adjectives obligatorily take the bound linker to combine with the head noun.

(8) a.   rìigaa baƙaa
           gown.f black.f

           'black gown'

b.   baƙa-r   rìigaa
black.f-L.f gown.f

'black gown/blackness of the gown'

(9)   a.   gidaa   baƙii
house.m black.m

'black house'

b.   baƙi-n   gidaa
black.m-L.m house.m

'black house/blackness of the house'

In contrast to possessives, pre-nominal adjectival constructions obligatorily make use of the bound linker: in other words, use of the free form linker is illicit here.[3]

(10)   a.   # baƙaa ta rìigaa
black.f L.f gown.f

'black gown'

b.   # baƙii   na   gidaa
black.m L.m house.m

'black house'

Superficially, pre-nominal adjectives resemble head nouns in N-of-N constructions. Semantically, though, they are clearly modifiers, just like their post-nominal counterparts. However, pre-head adjuncts are otherwise quite exotic in Hausa, which is a strict head-initial language. If we assume that pre-nominal adjectives are syntactic heads, but modifiers semantically, we shall be able to assimilate the analysis of this particular construction to the general word order of the language.

1.4   Arguments against Cliticisation

In her dissertation, Tuller (1986) treats na/ta-insertion analogously to the then-current GB account of English of-insertion, namely as a "dummy" case marker. Furthermore, she treats alternation between free form and bound form linker essentially as an instance of cliticisation, suggesting that both bound and free forms are syntactically part of the dependent NP. Phonologically, however, the bound marker is assumed to attach to the preceding N or N' node.[4]

   While sharing some initial plausibility, there are nevertheless a number of facts that make the surface cliticisation approach unmaintainable in the long run.

---

[3] Many Hausa adjectives have a second reading as an abstract noun. Thus, *farin gidaa* is ambiguous between 'the white house' and 'the whiteness of the house'. In the following discussion, I shall use the hash sign to mark the unavailability of the intended adjectival reading.

[4] The phonological change of feminine singular *ta* to *-r* is derived by means of a rhotacism rule.

First, the "cliticised" and free linker differ as to their agreement pattern: as detailed in Newman (2000), choice of *na* vs. *ta* is subject only to morpho-syntactic properties, namely the number and gender of the NP the possessive attaches to. If the attachment site is feminine singular, the free form linker is *ta*, otherwise, it is *na* (masculine or plural). The bound linker, however, observes some additional phonological constraint: -*r* is only possible with feminine singular hosts ending in /a/. As a result, feminine singular head nouns not ending in /a/ take *ta* as a free linker, but -*n* as a bound linker. Note that [a]-final masculines still take the linker -*n* (or *na*), illustrating that the constraints operative for bound forms are not purely phonological.

(11) a. màataa ta Bellò
          mother L.f Bello

          'Bello's mother'

      b. màata-r   Bellò
          mother-L.f Bello

          'Bello's mother'

(12) a. gwamnatì ta Ingilà
          government L.f England

          'England's government'

      b. gwamnatìn   Ingilà
          government-L England

          'England's government'

(13) a. ɓeeraa na Audù
          rat   L.m Audu

          'Audu's rat'

      b. ɓeera-n Audù
          rat-L.m Audu

          'Audu's rat'

Thus, under a surface-phonological cliticisation account, these hosts do not combine with a free form linker from which the bound form could possibly be derived.[5]

Second, not all occurrences of the bound linker can be related to a free form by regular phonological rules. As for non-pronominal possessors, reduction of *na/ta* to -*n*/-*r* can be modelled on the basis of a chain of fairly general rules, involving deletion of the short vowel (reduction), resyllabification as the coda of the preceding word, automatic shortening of the preceding vowel and finally, in the case of *ta*, rhotacism. With possessive pronouns, which, by contrast, feature a long vowel linker, plausibility of an automatic vowel deletion rule is greatly reduced.

---

[5] The criticism raised here does not, of course, apply to accounts within Distributional Morphology (Halle and Marantz, 1993).

(14)   Paradigm of free possessive pronouns

	masc/plural		feminine singular	
	sg	pl	sg	pl
1	nàawaa	naamù	tàawaa	taamù
2m	naakà	naakù	taakà	taakù
2f	naakì	naakù	taaki	taakù
3m	naasà/naashì	naasù	taasà	taasù
3f	naatà	naasù	taatà	taasù

(15)   Paradigm of bound genitive pronouns

	masc/plural		feminine singular	
	sg	pl	sg	pl
1	-naa	-nmù	-taa	-rmù
2m	-nkà	-nkù	-rkà	-rkù
2f	-nkì	-nkù	-rki	-rkù
3m	-nsà/-nshì	-nsù	-rsà/-rshì	-rsù
3f	-ntà	-nsù	-rsà	-rsù

Yet even if we were to accept automatic vowel reduction to apply to the long vowel pronominal possessive linker, bound possessive pronouns still confront us with idiosyncratic morpho-phonological effects: most notably, first person singular possessives undergo exceptional reduction, involving deletion of intervocalic /w/. Furthermore, first singular pronouns, in contrast to the rest of the paradigm obligatorily trigger lengthening of the preceding syllable.

(16)   a.   àku
              parrot

              'parrot'

        b.   * àkuu
              parrot

(17)   a.   àkuu-naa
              parrot-my

              'my parrot'

        b.   * àku-naa
              parrot-my

Third, in the case of pre-nominal adjectives, no construction with a full form linker exists from which the bound form could possibly be derived. As described in subsection (1.3) above, constructions with post-nominal adjectives do not feature a linker, whether bound or free, whereas pre-nominal adjectives are obligatorily marked with a bound linker. Furthermore, use of the free form linker is illicit in this construction, in contrast to possessives. Tuller (1986) recognises this problem, suggesting that pre-nominal adjectival constructions are actually compounds. Regular N-of-N compounds like *gidan wayàa* 'post office (lit.: house of wire)' or *gidan sauroo* 'mosquito net (lit.: house of mosquito)' indeed pattern with pre-nominal adjectives in that both disallow the use of the free linker. Tuller's perspective appears to be further supported by the observation that some pre-nominal adjectival constructions have acquired idiomatic semantics: to name an example, *farin gidaa* ' white house' can be used to refer to the White House as well as to any white house, whereas post-nominal *gidaa farii* 'white house' only has the compositional meaning (Newman, 2000). Thus, if the compositional meaning is still available for pre-nominal adjectival constructions, Tuller's compound analysis is greatly weakened. Interestingly enough, Tuller herself cites the perfectly compositional cases given below (Tuller, 1986, p. 36):

(18)   a. ƙàrama-r rìigaa baƙaa
           small-L.f gown black

           'small black gown'

       b. baƙa-r   rìigaa ƙàramaa
           black-L.f gown   small

           'black small gown'

Tuller (1986) suggests that the genitive linker has been reanalysed as a compound marker here. This solution, which is driven by theory-internal considerations, does not seem entirely satisfactory: while there is a plausible lexicalisation path from entirely compositional N-of-N constructions to idiomatic N-of-N compounds, no such path exists for pre-nominal adjectives, unless we concede that compositional pre-nominal adjectival constructions are syntactic in the first place. Thus, I would like to conclude that a compound analysis of pre-nominal adjectives introduces more conceptual problems than it actually solves. Furthermore, a dual analysis of the bound linker as compound marker and case marker must be inferior on Occamian grounds to any unified account of the data. I shall return to the case of idiomatic N-of-N and A-of-N compounds in the course of my analysis.

Finally, the bound marker is highly selective with respect to its host, attaching to nominal expressions only. In particular, phonological cliticisation to an immediately preceding verb is impossible:

(19)   Naa   karàntà na   Kànde
       1.sg.completive read   L.m Kànde

       'I read Kande's.'

(20)   * Naa   karàntà-n/-r Kànde
       1.sg.completive read-L   Kànde

To summarise, we have presented phonological, morphological and syntactic evidence to the extent that a surface-phonological cliticisation analysis of the Hausa bound genitive linker cannot account for the full range of data. Instead, I shall conclude that the language actually has two marking strategies for genitival constructions — morphological head-marking and syntactic dependent-marking — which partially overlap in the case of possessives.

2 Analysis

In the preceding section we have established that at least one of the uses of bound genitive markers, namely strong verbal nouns are unmistakable syntactic and semantic heads, taking their internal semantic arguments as syntactic complements. Furthermore, we observed that pre-nominal adjectives are syntactically similar to heads in N-of-N constructions and that pre-head adjuncts are actually quite exotic in this language. I shall therefore suggest that constructions involving the bound genitive marker, i.e., head-marking constructions, are to be uniformly analysed as head-complement structures. Since some of the constructions, most notably possessives and pre-nominal adjectives, involve semantic modification rather than complementation, I shall suggest that Hausa adjectives and possessed nouns undergo operations similar to type-raising, reversing functor-argument relations.

As a first step towards a formal analysis of genitive head marking in Hausa, we need to address of course how these markers get introduced on their host. As I have argued above, there is considerable reason to doubt the viability of a cliticisation approach. Rather, I shall suggest that the bound linker is a suffix, attached to its host by means of inflectional rules.

The inflectional perspective on the bound linker is also supported by the observation that its use is highly grammaticalised. As we have seen above, the bound linker marks a heterogeneous set of constructions: except with possessives, the linker does not appear to add any semantic contribution. Thus, I shall suggest that the basic function of the bound linker is to mark noun-complement structures. Under this perspective, the bound genitive linker is actually the nominal counterpart of direct object marking already attested for verbs (Crysmann, 2005).

Along with most recent work in HPSG, I assume that lexical entries are lexemes, not words. Fully inflected words are derived from lexemes by means of inflectional rules which are modelled as unary rule schemata (cf. Riehemann, 1998; Koenig, 1999).

The most fundamental property of direct object marking is that it only applies with direct object in situ. Adapting the proposal advanced in Crysmann (2005), I shall suggest that all direct object marking rules will inherit from the following constraint, which states that direct object marking is restricted to lexemes subcategorising for a complement bearing structural case.

(21)   *d-o-m-word* →

$$
\begin{bmatrix}
word \\
\text{SS} \quad \boxed{1} \\
\text{MORPH} \quad \begin{bmatrix} lexeme \\ \text{SS} \quad \boxed{1} \begin{bmatrix} \text{L} \mid \text{CAT} \mid \text{COMPS} \left\langle \begin{bmatrix} \text{L} \mid \text{CAT} \mid \text{HD} \mid \text{CASE} \quad struc \end{bmatrix}, \ldots \right\rangle \end{bmatrix} \end{bmatrix}
\end{bmatrix}
$$

The inflectional rules attaching the genitive linker are then instances of this more general rule type. The inflectional rule for -r is illustrated below.[6]

$$
(22) \quad
\begin{bmatrix}
d\text{-}o\text{-}m\text{-}word \\
\text{PHON} \quad \boxed{0} \ \oplus \ \langle r \rangle \\
\text{MORPH} \quad
\begin{bmatrix}
lexeme \\
\text{PH} \quad \boxed{0} \langle ...,a \rangle \\
\text{SS | L | CAT | HD} \quad
\begin{bmatrix}
noun \\
\text{AGR} \quad
\begin{bmatrix}
\text{NUM} & sg \\
\text{GEND} & fem
\end{bmatrix}
\end{bmatrix}
\end{bmatrix}
\end{bmatrix}
$$

These two direct object marking rules are then complemented by a third rule that derives zero marking in all other cases:

$$
(23) \quad
\begin{bmatrix}
plain\text{-}word \\
\text{PHON} \quad \boxed{0} \\
\text{MORPH} \quad
\begin{bmatrix}
lexeme \\
\text{PH} \quad \boxed{0} \\
\text{SS | L | CAT | COMPS} \ \neg \ \langle \left[\text{L | CAT | HD | CASE} \quad struc \right], \ ... \rangle
\end{bmatrix}
\end{bmatrix}
$$

Bound genitive pronouns can subsequently be added by means of a word-to-word inflectional rule. Essentially the rules introducing bound pronominals from the genitive set are almost identical to those introducing bound accusative pronominals. Selection of genitive vs. accusative sets is captured by means of different restrictions regarding the host's head value (*noun* vs. *verb*).

As depicted below, pronominal affixation saturates a direct object valency, and introduces the pronoun semantics into the host's MRS representation. Building on Copestake et al. (2005), syncategorematic introduction of contents by unary rule application will be performed via the constructional content feature C-CONT.

---

[6] Following Kathol (1999) and Wechsler and Zlatić (2001), I assume a head feature AGR for morpho-syntactic concord that is distinct from syntacto-semantic INDEX features. The main motivation in Hausa comes from the fact that formal agreement is marked even on dynamic nouns and nominal gerund, where INDEX actually refers to the event variable.

$$(24) \quad \begin{bmatrix} word \\ \text{PH} \quad \boxed{p} \ \oplus \ \langle \text{kì} \rangle \\ \\ \text{C-CONT} \ \boxed{c} \ \begin{bmatrix} \text{RELS} \ \left\langle \begin{bmatrix} \text{PRED} \quad "pron\text{-}rel" \\ \text{LBL} \quad \boxed{l} \\ \text{ARG0} \ \boxed{i} \begin{bmatrix} \text{PER} & 2 \\ \text{NUM} & sg \\ \text{GEND} & fem \end{bmatrix} \end{bmatrix} , \begin{bmatrix} \text{PRED} \quad "quant\text{-}rel" \\ \text{ARG0} \ \boxed{i} \\ \text{RSTR} \ \boxed{h} \end{bmatrix} \right\rangle \\ \\ \text{HCONS} \ \left\langle \begin{bmatrix} qeq \\ \text{HARG} \ \boxed{h} \\ \text{LARG} \ \boxed{l} \end{bmatrix} \right\rangle \end{bmatrix} \\ \\ \text{SS} \quad \begin{bmatrix} \text{L} \mid \text{CAT} \begin{bmatrix} \text{HEAD} \quad noun \\ \text{COMPS} \ \boxed{r} \end{bmatrix} \end{bmatrix} \\ \\ \text{MORPH} \ \begin{bmatrix} word \\ \text{PH} \quad \boxed{p} \\ \text{SS} \quad \begin{bmatrix} \text{L} \mid \text{CAT} \mid \text{COMPS} \left\langle \begin{bmatrix} \text{L} \mid \text{CONT} \ \boxed{c} \end{bmatrix} \right\rangle \oplus \boxed{r} \end{bmatrix} \end{bmatrix} \end{bmatrix}$$

Now that we have sketched how exponents of genitival direct object marking will be introduced onto their hosts, we can proceed towards a detailed account of the three constructions under consideration here which are instances of this general marking strategy.

2.1   Strong Verbal Nouns

The major difference between strong verbal nouns and their corresponding verbs is almost exclusively morphological in nature: derivation of a strong verbal noun from a verb base essentially changes the major category, but carries over unmodified the argument structure. As a result, grade 2 verbs display direct object marking of the verbal type, including final vowel shortening (Hayes, 1990; Crysmann, 2005), grade 2 strong verbal nouns, being nouns make use of nominal direct object marking, the genitive linker. Similarly, grade 2 verbs take bound object pronouns from the accusative set, whereas the derived strong verbal nouns select their genitive counterparts.

In order to account for strong verbal nouns, all we need to do is to provide a lexeme-to-lexeme rule that converts a verbal lexeme into a nominal one. Since verbal nouns can be specified for continuative aspect, I shall suggest that the semantic representation is equally carried over from the verb it derives from. In other words, strong verbal nouns are morpho-syntactic nouns with essentially verbal semantics.

(25)

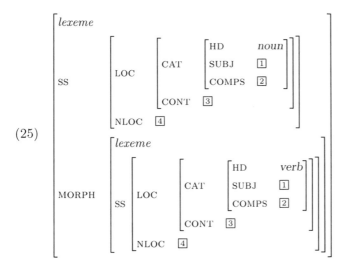

As we have observed above, direct object marking only applies, if the object is realised in situ. This is true not only for strong verbal nouns (see above), but also for plain verbs.

(26) a. rìigaa cèe   ya   dinkàa
         gown FOCUS 3.sg.m.completive sew

         'He has sewn.'

      b. Yaa   dinkà rìigaa
         3.sg.m.completive sew   gown

         'He has sewn the gown.'

It appears, thus, that extracted complements are simply not visible to direct object marking inflection.

The standard treatment of complement extraction in HPSG is trace-less (Pollard and Sag, 1994, ch. 9), that is the locally unexpressed valency is removed from the COMPS list and its *local* value is inserted directly into SLASH.

(27)

$$
\begin{bmatrix}
lexeme \\
\text{SS} \begin{bmatrix}
\text{LOC} \begin{bmatrix}
\text{CAT}\,|\,\text{COMPS} & \boxed{r} \\
\text{CONT}\,|\,\text{HOOK}\,|\,\text{INDEX} & event
\end{bmatrix} \\
\text{NLOC} \begin{bmatrix} \text{SLASH} & \{\boxed{l}\} \cup \boxed{s} \end{bmatrix}
\end{bmatrix} \\
\text{MORPH} \begin{bmatrix}
lexeme \\
\text{SS} \begin{bmatrix}
\text{LOC} \begin{bmatrix} \text{CAT}\,|\,\text{COMPS} & \left\langle \begin{bmatrix} \text{LOC} & \boxed{l} \\ \text{NLOC}\,|\,\text{SLASH} & \{\boxed{l}\} \end{bmatrix} \right\rangle \circ \boxed{r} \end{bmatrix} \\
\text{NLOC} \begin{bmatrix} \text{SLASH} & \boxed{s} \end{bmatrix}
\end{bmatrix}
\end{bmatrix}
\end{bmatrix}
$$

For Hausa complement extraction, this can be achieved with the lexeme-to-lexeme rule given above which generalises across verbs, verbal nouns and action nouns. If complement extraction is a lexeme-to-lexeme rule, applying before any lexeme-to-word rules, it is clear that in the case of extraction the direct object valency will have disappeared from COMPS at the point where the inflectional rules apply. As a result, the standard treatment of extraction in HPSG provides a straightforward account of the direct object marking patterns found in this language (cf. Crysmann, 2005 for a similar proposal).

2.2   Possessives

In contrast to verbal nouns and pre-nominal adjectives where the genitive linker does not contribute to the semantics, the bound possessive linker appears to be nothing more than the affixal counterpart of free form *na/ta*. However, we have established above that the bound form cannot be derived form the free form by simple cliticisation, so we are clearly dealing with an instance of head-marking here. While head-marking for in-situ direct objects is common in Hausa, it is otherwise unattested for head-adjunct structures. Furthermore, the morphological constraints applying to the bound marker are actually identical across all three constructions, suggesting, again, that we are in fact dealing with the same marker. I shall therefore propose that even in the possessive construction, the bound linker *-n/-r* is a direct object marker. In contrast to the free form, where the possessive relation is directly encoded in the lexical entry of *na/ta*, possessive semantics is introduced as part of the type raising rule that turns the possessor into the head noun's complement.

$$
(28)\quad
\begin{bmatrix}
\textit{lexeme} \\[2pt]
\text{C-CONT} \begin{bmatrix} \text{RELS} \left\langle \begin{bmatrix} \text{PRED} & \textit{"poss-rel"} \\ \text{LBL} & \boxed{l} \\ \text{ARG1} & \boxed{1} \\ \text{ARG2} & \boxed{2} \end{bmatrix} \right\rangle \\ \text{HCONS} \ \langle\rangle \end{bmatrix} \\[2pt]
\text{SS} \begin{bmatrix} \text{L} \begin{bmatrix} \text{CAT} \begin{bmatrix} \text{HD} & \boxed{0}\ \textit{noun} \\ \text{COMPS} & \left\langle \begin{bmatrix} \text{L} \mid \text{CONT} \mid \text{HOOK} \mid \text{INDEX} & \boxed{2} \end{bmatrix} \right\rangle \end{bmatrix} \\ \text{CONT} \mid \text{HOOK} \begin{bmatrix} \text{LTOP} & \boxed{l} \\ \text{INDEX} & \boxed{1} \end{bmatrix} \end{bmatrix} \end{bmatrix} \\[2pt]
\text{MORPH} \begin{bmatrix} \textit{lexeme} \\ \text{SS} \mid \text{L} \begin{bmatrix} \text{CAT} \begin{bmatrix} \text{HD} & \boxed{0} \\ \text{COMPS} & \langle\rangle \end{bmatrix} \end{bmatrix} \end{bmatrix}
\end{bmatrix}
$$

2.3   Adjectives

The treatment of pre-nominal vs. post-nominal attributive adjectives finally turns out to follow pretty much directly from our proposal as developed thus far.

Before we enter into a formal treatment of Hausa adjectives, a few words are due regarding the categorial status of these elements. Morphologically, they are very similar to nouns, being almost indistinguishable. To underline this property, Parsons (1963) introduced the term "dependent noun". Newman (2000) therefore suggests that adjectives in Hausa should rather be defined syntactically and semantically, being modifiers inherently. Furthermore, in contrast to nouns, they do not have inherent gender, but only agreement gender. In a theoretical framework like HPSG, however, which assumes syntactic categories to be complex feature structures, morphological, syntactic and semantic properties can be described independently, obviating the need to assign this class of lexeme a distinct head type of its own. Thus, I propose that basic adjectives in Hausa are nominal modifiers lacking inherent gender. Given their morphological similarity to other nominal expressions, I shall assume their HEAD value to be of type *noun* as well.

Thus, basic post-nominal adjectives will have a lexical entry along the following lines:

(29)

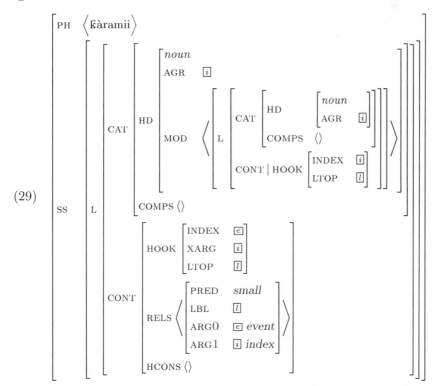

As depicted above, attributive adjectives in Hausa are nouns which select the N'-constituent they modify via MOD. With the exception of the categorial specification *noun*, the representation suggested here conforms to the standard treatment of attributive adjectival modifiers in HPSG as proposed for English in Pollard and Sag (1994). Equally unremarkable is the specification of intersective modifier semantics, which is essentially that proposed in Copestake et al. (2005). Agreement with the head noun is enforced by structure-sharing of the adjectives AGR value with the INDEX (and AGR) of the modified head noun.

Since basic adjectives do not take any direct object complements themselves, their COMPS list is empty. Thus, with respect to direct object marking, they obligatorily undergo the zero marking rule.

Given that Hausa is head initial and, moreover, that basic adjectives are adjuncts, they obligatorily follow the head they modify. Pre-nominal adjectives are then derived from basic adjectives by a lexical type raising rule, moving the selectional requirement towards the noun from MOD to COMPS.

$$
(30) \quad
\begin{bmatrix}
lexeme \\
\text{SS} \mid \text{L}
\begin{bmatrix}
\text{CAT}
\begin{bmatrix}
\text{HD}
\begin{bmatrix}
noun \\
\text{AGR} & \boxed{i} \\
\text{MOD} & \langle\rangle
\end{bmatrix} \\
\text{COMPS} \; \langle \boxed{m} \rangle
\end{bmatrix} \\
\text{CONT} \; \boxed{c}
\end{bmatrix} \\
\text{MORPH}
\begin{bmatrix}
word \\
\text{SS} \mid \text{L}
\begin{bmatrix}
\text{CAT}
\begin{bmatrix}
\text{HD}
\begin{bmatrix}
noun \\
\text{AGR} & \boxed{i} \\
\text{MOD} & \langle \boxed{m} \rangle
\end{bmatrix} \\
\text{COMPS} \; \langle\rangle
\end{bmatrix} \\
\text{CONT} \; \boxed{c}
\end{bmatrix}
\end{bmatrix}
\end{bmatrix}
$$

Since the semantics of adjectives is already fixed at the level of the lexical entry, the reversal of head-dependent relation can straightforwardly account for the alternation in word order and direct object marking, while leaving the semantic relation entirely unaffected.

3   Conclusion

Hausa bound genitive markers are used in a variety of grammatical constructions including strong verbal nouns, possessives, and pre-nominal adjectives. Despite the superficial difference in function, the marker is subject to identical morphological constraints, setting it apart from the free form possessive linker. Using type raising rules for bound possessive and pre-nominal adjectival constructions, we have developed a unified account of the genitive marker that not only captures the shared properties across its different uses, but also connects this head-marking strategy to a salient feature of the language, namely direct object marking.

References

Copestake, A., Flickinger, D., Pollard, C., Sag, I.: Minimal recursion semantics: an introduction. Research on Language and Computation 3(4), 281–332 (2005)

Crysmann, B.: An inflectional approach to Hausa final vowel shortening. In: Booij, G., van Marle, J. (eds.) Yearbook of Morphology 2004, pp. 73–112. Kluwer, Dordrecht (2005)

Halle, M., Marantz, A.: Distributed morphology and the pieces of inflection. In: Hale, K., Keyser, S.J. (eds.) The View from Building 20. Essays in Linguistics in Honor of Sylvain Bromberger. Current Studies in Linguistics, vol. 24, pp. 111–176. The MIT Press, Cambridge (1993)

Hayes, B.: Precompiled phrasal phonology. In: Inkelas, S., Zec, D. (eds.) The Phonology-Syntax Connection, pp. 85–108. University of Chicago Press, Chicago (1990)

Iida, M., Manning, C., O'Neill, P., Sag, I.: The lexical integrity of japanese causatives. Presented at the LSA 1994 Annual Meeting, Boston. Unpublished ms., Stanford University (1994)

Jaggar, P.: Hausa. John Benjamins, Amsterdam (2001)

Kathol, A.: Agreement and the syntax-morphology interface in HPSG. In: Levine, R., Green, G. (eds.) Studies in Contemporary Phrase Structure Grammar, pp. 209–260. Cambridge University Press, Cambridge (1999)

Kim, J.-B., Sag, I.: The parametric variation of French and English negation. In: Proceedings of WCCFL. CSLI publications, Stanford (1995)

Koenig, J.-P.: Lexical Relations. CSLI publications, Stanford (1999)

Newman, P.: The Hausa Language. An Encyclopedic Reference Grammar. Yale University Press, New Haven (2000)

Parsons, F.: The operation of gender in Hausa: Stabilizer, dependent nominals and qualifiers. African Language Studies 4, 166–207 (1963)

Parsons, F.W.: The verbal system in Hausa. Afrika und Übersee 44, 1–36 (1960)

Pollard, C., Sag, I.: Head–Driven Phrase Structure Grammar. CSLI and University of Chicago Press (1994)

Riehemann, S.: Type-based derivational morphology. Journal of Comparative Germanic Linguistics 2, 49–77 (1998)

Tuller, L.A.: Bijective Relations in Universal Grammar and the Syntax of Hausa. PhD thesis, UCLA, Ann Arbor (1986)

Wechsler, S., Zlatić, L.: A theory of agreement and its application to serbo-croatian. Language, 391–423 (2001)

The Generative Capacity of the Lambek–Grishin Calculus: A New Lower Bound

Matthijs Melissen

Universiteit Utrecht, The Netherlands
Université de Luxembourg, Luxembourg
info@matthijsmelissen.nl
http://www.matthijsmelissen.nl

Abstract. The Lambek–Grishin calculus **LG** is a categorial type logic obtained by adding a family of connectives $\{\oplus, \oslash, \obslash\}$ dual to the family $\{\otimes, /, \backslash\}$, and adding interaction postulates between the two families of connectives thus obtained. In this paper, we prove a new lower bound on the generative capacity of **LG**, namely the class of languages that are the intersection of a context-free language and the permutation closure of a context-free language. This implies that **LG** recognizes languages like the MIX language, e.g. the permutation closure of $\{a^n b^n c^n \mid n \in \mathbb{N}\}$, and $\{a^n b^n c^n d^n e^n \mid n \in \mathbb{N}\}$, which can not be recognized by tree adjoining grammars.

1 Introduction

This paper studies the weak generative capacity, e.g. the class of languages that can be recognized, of the *Lambek–Grishin calculus* (**LG**) proposed by Moortgat [12]. We prove that any intersection of a context-free language and the permutation closure of a context-free language can be generated by **LG**.

The Lambek–Grishin calculus is an extension of the nonassociative Lambek calculus **NL** [10], obtained by allowing for structural product connectives on the right-hand side of sequents, and adding interaction principles between the families of connectives that correspond to the left-hand side and right-hand products. For **NL**, it has been proved by Kandulski [9] that this calculus is weakly equivalent to the context-free grammars. The same result for the associative variant of this calculus, the associative Lambek calculus **L**, has been proved by Pentus [15]. Context-free languages are, however, generally assumed to be not expressive enough for natural language [6]. Moot [13] has proved that any *tree adjoining grammar* [7] can be converted into an equivalent **LG**-grammar, which implies that **LG** is at least *mildly context-sensitive*. Therefore **LG** seems to be more suitable for natural language. The exact generative capacity of **LG** is currently unknown. Moot [13] also proposed a proof for conversion of **LG** into tree adjoining grammar, but this proof does not apply to the general case, as the current paper shows.

The language $\{a^n b^n c^n d^n e^n \mid n \in \mathbb{N}\}$ cannot be generated by tree adjoining grammars [8], but we will show that this language can in fact be generated by

P. de Groote, M. Egg, and L. Kallmeyer (Eds.): FG 2009, LNAI 5591, pp. 118–132, 2011.

LG. On the other hand, the COPY language $\{ww|w \in \{a, b\}^*\}$ can be generated by a tree adjoining grammar, but is not element of the lower bound proved in this paper. Therefore neither the lower bound proved by Moot (the languages generated by tree adjoining grammars) is a subset of the lower bound proved in this paper (the class of languages that are the intersection of a context-free language and the permutation closure of a context-free language), nor the other way around. This means that the current known lower bound on the generative capacity of **LG** is the union of those two classes. The only currently known upper bound on the generative capacity of **LG** is decidability [12], which makes the class of languages generated by this calculus at most recursive.

After the introduction, we first give a definition of the Lambek–Grishin calculus in Sect. 2. Then we proceed by proving the new lower bound on the generative capacity in Sect. 3. In Sect. 4 we give some examples of languages that can be generated, and in Sect. 5 some concluding remarks will be made.

2   The Lambek–Grishin Calculus

The *Lambek–Grishin calculus* **LG** is an extension of the non-associative Lambek calculus **NL**. In **NL** there are only product operators in a structural role on the left-hand side of the sequent, while in **LG**, such connectives are also allowed on the right-hand side. We have therefore two different families of connectives, one related to each of both products. **LG** is a deductive system. A sentence is grammatical if and only if there exists a sequent belonging to this sentence that is derivable in the deductive system.

Definition 1. *Sequents in **LG** have the form $p \rightarrow q$ where p and q are types. We assign to every word in the language a finite number of types. We have a set of basic types called Atoms. The set of types is defined as follows: every atom is a type, and if p and q are types, then the following formulas are types as well:*

$$p \otimes q, \quad p \backslash q, \quad p/q, \quad p \oplus q, \quad p \oslash q, \quad p \oslash q.$$

*An axiomatization of **LG** is displayed in Table 1. We call $\otimes$ and $\oplus$ product connectives, and $\backslash$, $/$, $\oslash$ and $\oslash$ implicational connectives. We say that in $q\backslash p$, p/q, $q \oslash p$ and $p \oslash q$, p appears* above *and q appears* under *the implication.*

We can see $\otimes$ as the product connective playing a structural role on the left-hand side of the arrow, and $\oplus$ as the product connective playing a structural role on the right-hand side. Intuitively one or more words with type $a\backslash b$ can be seen as a group of words which need a phrase of type a on their left, and return a phrase of type b. In the same way we can see a phrase of type a/b as a phrase which needs a phrase of type b on its right, before returning an a type phrase. Finally the $\otimes$ operator can be seen as concatenation of words. The connectives $\oplus$, $\oslash$ and $\oslash$ (pronounced respectively plus, from and minus) are the duals of $\otimes$, $/$ and $\backslash$ under arrow reversal, i.e. the symbol $\oplus$ fulfills the same role on the right-hand side of the arrow as the symbol $\otimes$ does on the left-hand side, and vice versa. Between the other connectives, similar relations hold.

Table 1. Axioms and rules of Lambek–Grishin calculus

$$t \to t \quad \textbf{Id} \quad \frac{b \to a\backslash c}{a \otimes b \to c} \; \textbf{Res} \quad \frac{a \oslash c \to b}{c \to a \oplus b} \; \textbf{Res}$$

$$(t \in Atoms) \quad \frac{}{a \to c/b} \; \textbf{Res} \qquad \frac{}{c \oslash b \to a} \; \textbf{Res}$$

$$\frac{c \to a \quad b \to d}{a\backslash b \to c\backslash d} \; \textbf{Mon} \qquad \frac{d \to b \quad a \to c}{a/b \to c/d} \; \textbf{Mon} \qquad \frac{a \to c \quad b \to d}{a \otimes b \to c \otimes d} \; \textbf{Mon}$$

$$\frac{c \to a \quad b \to d}{a \oslash b \to c \oslash d} \; \textbf{Mon} \qquad \frac{d \to b \quad a \to c}{a \oslash b \to c \oslash d} \; \textbf{Mon} \qquad \frac{a \to c \quad b \to d}{a \oplus b \to c \oplus d} \; \textbf{Mon}$$

$$\frac{a \oslash (b \otimes c) \to d}{(a \oslash b) \otimes c \to d} \; \textbf{G}_1 \quad \frac{b \oslash (a \otimes c) \to d}{a \otimes (b \oslash c) \to d} \; \textbf{G}_2 \quad \frac{(a \otimes b) \oslash c \to d}{a \otimes (b \oslash c) \to d} \; \textbf{G}_3 \quad \frac{(a \otimes c) \oslash b \to d}{(a \oslash b) \otimes c \to d} \; \textbf{G}_4$$

There are multiple equivalent axiomatizations for **LG**, of which one is displayed in Table 1. A double line denotes derivability in two directions. We assume loop detection, i.e. we exclude repetitions within derivations. This axiomatization is particularly interesting because it is cut-free and allows for decidable proof search. It uses only logical and no structural connectives; the role of the products is played by $\otimes$ on the left-hand side and $\oplus$ on the right-hand side. We have *identity* (**Id**) as an axiom (where t is an atom). Furthermore there are two pairs of *residuation* (**Res**) rules. The left pair is the same as in **NL**, and the right pair is the symmetric version of the left pair. We also have six *monotonicity rules* (**Mon**). Finally the *Grishin interactions* $\textbf{G}_1 - \textbf{G}_4$, based on a paper by Grishin [5], are added. Those postulates govern the interaction between the $\oplus$- and $\otimes$-families.

Lemma 1 ([12]). *Transitivity, i.e. $a \to b$ and $b \to c$ implies $a \to c$ is admissible in* **LG**.

Now we will define how we can use **LG** to determine what sentences are grammatical.

Definition 2. *An* **LG**-*grammar has the form $\mathcal{L}(\Sigma, s, \varphi)$, where Σ is a finite alphabet, s the goal type (i.e. the type corresponding to an entire sentence) and φ a mapping, called the type dictionary, that assigns one or more types to every word. We require that for all $t \in \Sigma$, $\varphi(t)$ is finite. The language generated by an* **LG**-*grammar $\mathcal{L}(\Sigma, s, \varphi)$ is defined as the set of all expressions $t_1 \ldots t_n$ over the alphabet Σ for which there exists a derivable sequent $b_1 \otimes \ldots \otimes b_n \to s$ (with some binary bracketing imposed on $b_1 \otimes \ldots \otimes b_n$) such that $b_i \in \varphi(t_i)$ for all $1 \leq i \leq n$.*

Example 1. Now we give an example derivation of the sentence 'Alice thinks someone left'. The resulting formula on the upper left is an **NL**-derivable sequent. For clarity, the atoms occurring in the type of 'someone' are typeset in bold.

$$
\cfrac{
\cfrac{
\cfrac{
\cfrac{
\cfrac{
\cfrac{
\cfrac{np \otimes (((np\backslash s)/s) \otimes (\mathbf{np} \otimes (np\backslash s))) \to \mathbf{s} \qquad \mathbf{s} \to s}{(np \otimes (((np\backslash s)/s) \otimes (\mathbf{np} \otimes (np\backslash s)))) \oslash s \to \mathbf{s} \oslash s} \ \text{Mon}
}{np \otimes (((np\backslash s)/s) \otimes (\mathbf{np} \otimes (np\backslash s))) \to (\mathbf{s} \oslash s) \oplus s} \ \text{Res}
}{(\mathbf{s} \oslash \mathbf{s}) \oslash (np \otimes (((np\backslash s)/s) \otimes (\mathbf{np} \otimes (np\backslash s)))) \to s} \ \text{Res}
}{np \otimes ((\mathbf{s} \oslash \mathbf{s}) \oslash (((np\backslash s)/s) \otimes (\mathbf{np} \otimes (np\backslash s)))) \to s} \ \mathbf{G_2}
}{np \otimes (((np\backslash s)/s) \otimes ((\mathbf{s} \oslash \mathbf{s}) \oslash (\mathbf{np} \otimes (np\backslash s)))) \to s} \ \mathbf{G_2}
}{np \otimes (((np\backslash s)/s) \otimes (((\mathbf{s} \oslash \mathbf{s}) \oslash \mathbf{np}) \otimes (np\backslash s))) \to s} \ \mathbf{G_1}
$$

$$
\underbrace{np}_{\text{Alice}} \otimes (\underbrace{((np\backslash s)/s)}_{\text{thinks}} \otimes (\underbrace{((\mathbf{s} \oslash \mathbf{s}) \oslash \mathbf{np})}_{\text{someone}} \otimes \underbrace{(np\backslash s)}_{\text{left}})) \to s
$$

Definition 3. *The product-free non-associative Lambek calculus or product-free* **NL** *can be obtained from the definition of* **LG** *by requiring that* / *and* \ *are the only connectives in the type dictionary.*

Note that in product-free **NL**, we can leave out all rules containing the Grishin connectives $\oplus$, $\oslash$ and $\oslash$.

Definition 4. *Spinal* **AB**-*grammar* (**AB$_s$**) *is a restricted categorial grammar where we only allow types of the form* a *and* $a\backslash b$, *where* a *and* b *are atomic types. The only rule is that we can replace any subexpression of the form* $a, a\backslash b$ *by* b *in the left-hand side of the arrow.*

It can easily be seen that all derivable sequents in **AB$_s$** have the form $(\dots(a_1, (a_1\backslash a_2)), \dots), (a_{n-1}\backslash a_n) \to a_n$. This calculus can be seen as a restricted version of the calculus used in Ajdukiewicz–Bar-Hillel grammar [1].

Lemma 2. *Any* **AB$_s$**-*sequent is an* **LG**-*sequent when we replace comma by* $\otimes$.

Proof. Note that the rule of **AB$_s$** can only be applied in positive context (i.e. under an even number of implicational connectives) in **AB$_s$**, namely under 0 implicational connectives in the left-hand side of a sequent. We can easily check that the application of the rule of **AB$_s$** in positive context in the left-hand side of a sequent is valid in **LG**.

Definition 5. *The language generated by a product-free* **NL**-*grammar is defined in the same way as the language generated by an* **LG**-*grammar; the language generated by an* **AB$_s$**-*grammar is also defined in the same way, except that we allow for an arbitrary number of goal types, instead of just one goal type.*

Definition 6. *We write* $a \div b$ *as an expression which may stand for both* a/b *or* $b\backslash a$. *A subformula is positive resp. negative if it occurs under an even resp. odd number of implicational connectives. The count of an atom in a type is the number of positive occurrences of this atom in the type minus the number of negative occurrences of this atom in the type, and the count of a set of atoms in a type is the sum of the counts of those atoms.*

Lemma 3 (*Count invariant*, e.g. [17]). *For all sets of atoms A, if $p \to q$ is derivable in **LG**, the count of A in p equals the count of A in q.*

3   Main Proof

Theorem 1. *Lambek–Grishin calculus is capable of recognizing any language that is the intersection of a context-free language and the permutation closure of a context-free language.*

The proof will follow at the end of this section. Because we require in **LG** non-empty left-hand sides in derivations, this calculus cannot generate the empty string, so we will ignore the recognition of the empty string in the equivalence proof.

Definition 7. *A deterministic finite automaton (DFA) (e.g. [16]) is a 5-tuple $\langle Q, \Sigma, \delta, q_0, F \rangle$, where Q is a finite set of states, Σ is a finite set of symbols, $\delta : Q \times \Sigma \to Q$ is the transition function, $q_0 \in Q$ is the initial state and $F \subseteq Q$ are the accept states. A string $\mathsf{a}_1 \ldots \mathsf{a}_n$ can be accepted by the DFA if there is a sequence of states $q_0, \ldots, q_n$ such that $q_n \in F$ and $\delta(q_{i-1}, \mathsf{a}_i) = q_i$ for $1 \leq i \leq n$.*

Example 2. Consider the following DFA:

$$\langle \{q_0, q_1, q_2, q_3, q_4\}, \{\mathsf{a}, \mathsf{b}, \mathsf{c}, \mathsf{d}\}, \delta, q_0, \{q_0, q_3\} \rangle$$

with δ as follows: $\delta(q_0, \mathsf{a}) = q_0$; $\delta(q_0, \mathsf{b}) = q_1$; $\delta(q_1, \mathsf{c}) = q_2$; $\delta(q_2, \mathsf{d}) = q_3$; $\delta(q_3, \mathsf{b}) = q_1$; for any other state q and input symbol i, $\delta(q, \mathsf{i}) = q_4$. This automaton recognizes an arbitrary number of a's, followed by an arbitrary number of sequences bcd. Note that q_4 acts here as a *sink*: it is not possible to leave this state.

Lemma 4. *For every regular language there is an equivalent $\mathbf{AB_s}$-language.*

Proof. It is well-known that the class of regular languages is exactly the class of languages recognized by DFA. We will show that any DFA can be simulated by an $\mathbf{AB_s}$-grammar, thereby proving our lemma. Our proof is similar to the proof in [17], where it is proved that **LP**, the Lambek calculus with permutation, recognizes exactly all permutations of context-free languages. However, in that proof the definition of regular languages is used, while we make use of the machine model.

Given a DFA $\langle Q, \Sigma, \delta, q_0, F \rangle$, we define an $\mathbf{AB_s}$-grammar with symbols Σ', goal types D and type dictionary φ as follows.

- $\Sigma' = \Sigma$;
- $D = F$;
- φ is defined as follows: $\varphi(\mathsf{a}) = \{q_1 \backslash q_2 \mid \delta(q_1, \mathsf{a}) = q_2\} \cup \{q \mid \delta(q_0, \mathsf{a}) = q\}$.

In other words, if we can move with symbol a from state q_1 to q_2, we assign to a the type $q_1 \backslash q_2$, and if we can immediately reach q from the initial state with symbol a, we give a type q.

If a string $t_1 \ldots t_n$ of length at least 1 can be accepted by the DFA, it means there is a sequence of states $q_0, \ldots, q_n$ such that $q_n \in F$ and $\delta(q_{i-1}, t_i) = q_i$ for $1 \leq i \leq n$. By our conversion, this means that $q_n \in D$, $q_1 \in \varphi(t_1)$ and $q_i \backslash q_{i+1} \in \varphi(t_{i+1})$ for $1 \leq i \leq n$. Because $q_1, q_1 \backslash q_2, \ldots, q_{n-1} \backslash q_n \rightarrow q_n$ is derivable in $\mathbf{AB_s}$, the string $t_1 \ldots t_n$ can be recognized by the $\mathbf{AB_s}$-grammar.

If $t_1 \ldots t_n$ is recognized by an $\mathbf{AB_s}$-grammar, it means that there exists an $\mathbf{AB_s}$-derivable sequent $b_1, \ldots, b_n \rightarrow d$ such that $b_i \in \varphi(t_i)$ for all $1 \leq i \leq n$ and $d \in D$. Because all $\mathbf{AB_s}$-derivations have the form $q_1, q_1 \backslash q_2, \ldots, q_{n-1} \backslash q_n \rightarrow q_n$, it holds that $q_n = d$, $q_1 \in \varphi(t_1)$ and $q_{i-1} \backslash q_i \in \varphi(t_i)$ for $2 \leq i \leq n$. By definition of φ, $\delta(q_0, t_1) = q_1$ and $\delta(q_{i-1}, t_i) = q_i$ for $2 \leq i \leq n$. By $q_n = d$ and $d \in D$, $q_n \in D$ so also $q_n \in F$. Now we know that there is a sequence of states $q_0, \ldots, q_n$ such that $q_n \in F$ and $\delta(q_{i-1}, a_i) = q_i$ for $1 \leq i \leq n$, so $t_1, \ldots, t_n$ can be recognized by the DFA. $\qquad\square$

Example 3. Consider the DFA from Example 2. We will create an $\mathbf{AB_s}$-language equivalent to this DFA according to the procedure of Lemma 4. This gives us symbols $\Sigma = \{a, b, c, d\}$, goal types $d = \{q_0, q_3\}$, and the following type dictionary φ:

p	$\varphi(p)$
a	$q_0, q_0 \backslash q_0, q_1 \backslash q_4, q_2 \backslash q_4, q_3 \backslash q_4, q_4 \backslash q_4$
b	$q_1, q_0 \backslash q_1, q_1 \backslash q_4, q_2 \backslash q_4, q_3 \backslash q_1, q_4 \backslash q_4$
c	$q_4, q_0 \backslash q_4, q_1 \backslash q_2, q_2 \backslash q_4, q_3 \backslash q_4, q_4 \backslash q_4$
d	$q_4, q_0 \backslash q_4, q_1 \backslash q_4, q_2 \backslash q_3, q_3 \backslash q_4, q_4 \backslash q_4$

To show that the string aaabcd is in the language, it suffices to prove

$$q_0, q_0 \backslash q_0, q_0 \backslash q_0, q_0 \backslash q_1, q_1 \backslash q_2, q_2 \backslash q_3 \rightarrow q_3 \ ,$$

which is indeed derivable in $\mathbf{AB_s}$. Of course, we could simplify the type dictionary by leaving out the types with the sink state q_4 in the right hand side of the $\backslash$.

Lemma 5. *For any product-free $\mathbf{NL}$-language $\mathcal{L}_1$ and $\mathbf{AB_s}$-language $\mathcal{L}_2$ there is an $\mathbf{LG}$-grammar that recognizes exactly the intersection of $\mathcal{L}_1$ and the permutation closure of $\mathcal{L}_2$.*

Proof. Consider a language $\mathcal{L}_1$ generated by a product-free $\mathbf{NL}$-grammar $\mathcal{G}_1$ with symbols Σ_1, goal type d and type dictionary φ_1, and a language $\mathcal{L}_2$ generated by an $\mathbf{AB_s}$-grammar $\mathcal{G}_2$ with symbols Σ_2, goal types D and type dictionary φ_2. Note that we can easily convert an $\mathbf{AB_s}$-grammar with multiple goal types D into an equivalent $\mathbf{AB_s}$-grammar with a single goal type, by choosing a fresh atomic goal type g and turning the type dictionary φ into φ', where for every symbol t, $\varphi'(\mathsf{t}) = \varphi(\mathsf{t}) \cup \{a \backslash g \mid a \backslash d \in \varphi(\mathsf{t}), d \in D\} \cup \{g \mid d \in \varphi(\mathsf{t}), d \in D, d \text{ is atomic}\}$. Therefore we can assume without loss of generality that $D = \{d\}$ (i.e. we make sure that $\mathcal{G}_1$ and $\mathcal{G}_2$ have the same, single goal type), that d is atomic, that d

does not occur in φ_2 in the left-hand side of $\backslash$, and that the intersection of the atoms in the range of φ_1 and in the range of φ_2 is $\{d\}$. We pick an atom s neither occurring in the range of φ_1 nor in the range of φ_2, $s \neq d$. We define T_1 as the union of $\{s\}$ and the atoms in the range of φ_1 and T_2 as the atoms in the range of φ_2, both excluding d. Note that d functions as goal of the existing grammars $\mathcal{G}_1$ and $\mathcal{G}_2$, while s will function as goal of the new grammar $\mathcal{G}$.

Now we can create an **LG**-grammar that recognizes the intersection of $\mathcal{L}_1$ and the permutation closure of $\mathcal{L}_2$ as follows. We take as symbols $\Sigma_1 \cap \Sigma_2$, and choose s as goal type. The type dictionary φ is defined in the following way. Given $\mathsf{p} \in \Sigma$, and $\varphi_1(\mathsf{p})$ and $\varphi_2(\mathsf{p})$, we define $\varphi(\mathsf{p})$ as follows:

$$\varphi(\mathsf{p}) = \{(b \oslash a) \otimes c \mid c \in \varphi_1(p), a \backslash b \in \varphi_2(p)\}$$
$$\cup \{(a \oslash s) \otimes c \mid c \in \varphi_1(p), a \in \varphi_2(p), a \text{ is atomic}\} .$$

($\Rightarrow$) First we will show that any string in the intersection of $\mathcal{L}_1$ and the permutation closure of $\mathcal{L}_2$ is also in $\mathcal{L}$. The steps we apply are displayed in Fig. 1. When the bracketing of $\otimes$ is not important, we just leave out the brackets. Assume we have a product-free **NL**-grammar $\mathcal{G}_1$ and an **AB$_\mathbf{s}$**-grammar $\mathcal{G}_2$. We define $\mathcal{G}$ to be the **LG**-grammar obtained by the procedure described above. Assume we have a string $\mathsf{t}_1 \ldots \mathsf{t}_n$ that is recognizable by $\mathcal{G}_1$, and a permutation π on $[1..n]$ such that $\mathsf{t}_{\pi(1)} \ldots \mathsf{t}_{\pi(n)}$ is recognizable by $\mathcal{G}_2$. We will show that this permuted string is recognizable by $\mathcal{G}$ as well. If $\mathsf{t}_{\pi(1)} \ldots \mathsf{t}_{\pi(n)}$ is recognizable by $\mathcal{G}_2$, there must be an **AB$_\mathbf{s}$**-derivation for (1) in Fig. 1 (note that all **AB$_\mathbf{s}$**-derivable sequents have this form) such that $b_n = d$, $b_1 \in \varphi_2(\mathsf{t}_{\pi(1)})$ and $b_{i-1} \backslash b_i \in \varphi_2(\mathsf{t}_{\pi(i)})$ for every i with $2 \leq i \leq n$. By Lemma 2, (2) is an **LG**-derivable sequent. By symmetry under arrow reversal, (3) is also derivable. By residuation and monotonicity in **LG**, we can derive (4). Because $\mathsf{t}_1 \ldots \mathsf{t}_n$ is recognizable by $\mathcal{G}_1$, we have a product-free **NL**-derivation and therefore also an **LG**-derivation for $a_1 \otimes a_2 \otimes \ldots, a_n \to d$ (with a binary bracketing imposed on the left-hand side) such that $a_i \in \varphi_1(\mathsf{t}_i)$ for every i with $1 \leq i \leq n$. Because $b_n = d$, we can obtain (5) by transitivity on this sequent and (4). Then we can move the $b_i \oslash b_{i-1}$-subformulas to the left-hand side by residuation, obtaining (6). Finally we move those types in the correct place with Grishin postulate $\mathbf{G}_1$. Note that we can place any $b_i \oslash b_{i-1}$ at any position under a_j, so we position them according to the permutation π, as depicted in (7), where π' is the inverse permutation of π. We now have a sequent of the form $(\ldots (c_1 \otimes c_2) \otimes \ldots) \otimes c_n \to b_0$ where b_0 is the goal type and each c_i has the form $(b_{\pi'(i)} \oslash b_{\pi'(i)-1}) \oslash a_i$. Because $b_1 \in \varphi_2(\mathsf{t}_{\pi(1)})$, $b_{i-1} \backslash b_i \in \varphi_2(\mathsf{t}_{\pi(i)})$ for $2 \leq i \leq n$ and $a_i \in \varphi_1(\mathsf{t}_i)$ for each $1 \leq i \leq n$, it holds that $c_i \in \varphi(\mathsf{t}_i)$ for $1 \leq i \leq n$ by definition of φ. Therefore $\mathsf{t}_1 \ldots \mathsf{t}_n$ is recognizable by $\mathcal{G}$. Now we can conclude that if a string can be recognized by a product-free **NL**-grammar and any permutation of it can be recognized by an **AB$_\mathbf{s}$**-grammar, the string can also be recognized by the corresponding **LG**-grammar obtained by the procedure described in this lemma.

($\Leftarrow$) Now we show that any string in $\mathcal{L}$ is both in $\mathcal{L}_1$ and the permutation closure of $\mathcal{L}_2$. We prove this by showing the reversibility of the steps in Fig. 1. Assume that $\mathsf{t}_1 \ldots \mathsf{t}_n$ is in $\mathcal{L}$. Then there is a sequent $d_1 \otimes \ldots \otimes d_n \to d$ (with

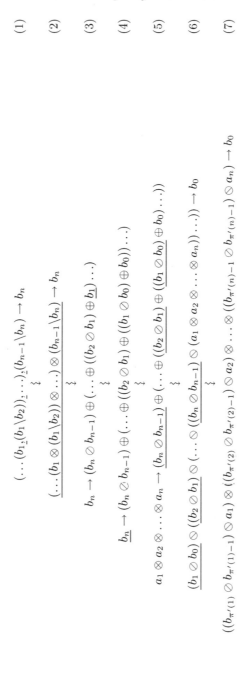

$$(\ldots(b_{1\underline{\cdot}}(b_1\backslash b_2))\underline{\cdot}\ldots)\underline{\cdot}(b_{n-1}\backslash b_n)\to b_n \tag{1}$$

$$(\ldots(b_1\otimes(b_1\backslash b_2))\otimes\ldots)\otimes\underline{(b_{n-1}\backslash b_n)}\to b_n \tag{2}$$

$$b_n\to(b_n\oslash b_{n-1})\oplus(\ldots\oplus((b_2\oslash b_1)\oplus\underline{b_1})\ldots) \tag{3}$$

$$\underline{b_n}\to(b_n\oslash b_{n-1})\oplus(\ldots\oplus((b_2\oslash b_1)\oplus((b_1\oslash b_0)\oplus b_0))\ldots) \tag{4}$$

$$a_1\otimes a_2\otimes\ldots\otimes a_n\to\underline{(b_n\oslash b_{n-1})}\oplus(\ldots\oplus(\underline{(b_2\oslash b_1)}\oplus(\underline{(b_1\oslash b_0)}\oplus b_0)\ldots)) \tag{5}$$

$$\underline{(b_1\oslash b_0)}\oslash(\ldots\oslash((\underline{b_n\oslash b_{n-1}})\oslash(a_1\otimes a_2\otimes\ldots\otimes a_n))\ldots))\to b_0 \tag{6}$$

$$((b_{\pi'(1)}\oslash b_{\pi'(1)-1})\oslash a_1)\otimes((b_{\pi'(2)}\oslash b_{\pi'(2)-1})\oslash a_2)\otimes\ldots\otimes((b_{\pi'(n)}\oslash b_{\pi'(n)-1})\oslash a_n)\to b_0 \tag{7}$$

Fig. 1. The stepwise transformation of an $\mathbf{AB_s}$-sequent into an $\mathbf{LG}$-sequent. In every line, the difference with the next line is underlined

some binary bracketing imposed on the left-hand side) and $d_i \in \varphi(\mathsf{t}_i)$ for every i. Sequents of this form will be called *initial*, which notion we will formalize in the next definition. Next, we will prove some properties about sequents of this form in the following lemmas, which we need to prove reversibility of (6)–(7). After that, we continue the current proof by proving reversibility of the other steps.

Definition 8. *Given two disjoint sets T_1, T_2 and $d \notin T_1 \cup T_2$, we define initial sequents as sequents $d_1, \ldots, d_n \to d$ such that each d_i has the form $(a_i \oslash b_i) \otimes c_i$ with $a_i \in T_2 \cup \{d\}$, $b_i \in T_2$ and $c_i \in T_1 \cup \{d\}$ (compare (7) in Fig. 1).*

In the following lemmas we will show that if an initial sequent can be derived, it can be done so by first moving all Grishin connectives outside the scope of any $\otimes$-connective (compare (6) in Fig. 1). In other words, we show reversibility of the step (6)–(7) in this figure. This is the crucial part of our proof.

Definition 9. *We define the following classes of types:*

$$\mathcal{F} ::= c \mid \mathcal{F} \div \mathcal{F} \; ;$$
$$\mathcal{P} ::= (a \oslash b) \otimes \mathcal{P} \mid \mathcal{F} \mid \mathcal{P} \otimes \mathcal{P} \; ;$$
$$\mathcal{Q} ::= \mathcal{F} \mid (a \oslash b) \oplus \mathcal{Q} \mid \mathcal{Q} \div \mathcal{P} \; ;$$
$$\mathcal{R} ::= a \mid (a \oslash b) \oplus \mathcal{R} \mid \mathcal{R} \div \mathcal{P} \; ;$$

such that $a \in T_2 \cup \{d\}$; $b \in T_2$; $c \in T_1 \cup \{d\}$, where a, b and c are atomic types, T_1 and T_2 are disjoint classes of arbitrary types and $d \notin T_1 \cup T_2$.

Note that $\mathcal{F}$ is the class of product-free Lambek types.

Lemma 6. *Each initial sequent has the form $p \to r$ with $p \in \mathcal{P}$ and $r \in \mathcal{R}$.*

Proof. This follows directly from Definition 8 (initial sequents). □

In other words, we can see sequents of the form $p \to r$ with $p \in \mathcal{P}$ and $r \in \mathcal{R}$ as a generalization of the class of initial sequents. As will become clear, the former class can be seen as the closure of the class of initial sequents under operations such as Grishin interactions and various forms of residuation.

We also need to adapt the idea of 'moving Grishin connectives outside the scope of any $\otimes$-connective' to this generalization. We replace this notion with the notion of *internal* Grishin connectives. Intuitively, a Grishin connective is internal if it does not occur within the scope of a $\otimes$-connective, taking a restricted number of forms of residuation into account. Now we give a formal definition.

Definition 10. *A subformula in a sequent $p \to r$ with $p \in \mathcal{P}$ and $r \in \mathcal{R}$ is internal if at least one of the folloing holds: (1) the subformula is in p while r contains $\div$; (2) the subformula is in the left argument of $\oplus$ in r while the right argument contains a $\div$-connective; (3) the subformula is within the scope of $\otimes$ in p; or (4) the subformula is in r in the right-hand argument of a $\div$-connective. Furthermore we say that a connective is internal if it occurs in an internal subformula.*

Lemma 7. *Consider the class of sequents $p \to q$ such that $p \in \mathcal{P}$ and $q \in \mathcal{Q}$ and either p or q (or both) contain one of the Grishin connectives $\oslash$, $\oslash$ and $\oplus$. Sequents in this class are underivable.*

Proof. We prove that all derivations of sequents in this class would be infinite, by showing that in order to prove a sequent $p \to q$ with $p \in \mathcal{P}$ and $q \in \mathcal{Q}$, we first need to prove another sequent that is also in this class. Because derivations need to have a finite length, no sequents in this class are derivable. Remember that the decidable axiomatization consists of the identity axiom for atomic types, residuation, monotonicity and the Grishin interactions in rule-based form. We proceed by case distinction on each of the four rules that could have been used to derive $p \to q$.

First note that $p \to q$ cannot be an axiom with atomic types, because the sequent is required to contain a connective.

Now we consider the case where the last proof step is monotonicity. Then the outer connectives on the left-hand and right-hand side must be equal. The only connectives that can appear as outer connective in both $\mathcal{P}$ and $\mathcal{Q}$ are $/$ and $\backslash$. In that case, p does not contain Grishin connectives and therefore q does, so we can write q as $q' \div p'$ with $p' \in \mathcal{P}$ and $q' \in \mathcal{Q}$ and p as $c_1 \div c_2$ with $c_1, c_2 \in \mathcal{F}$. Therefore the conclusion of the monotonicity rule has the form $c_1 \div c_2 \to q' \div p'$. Then we need to derive $c_1 \to q'$ and $p' \to c_2$. Because q contains Grishin connectives, either p' or q' (or both) contains Grishin connectives. If p' contains Grishin connectives, $p' \to c_2$ is again in the class and if q' contains Grishin connectives, $c_1 \to q'$ is again in the class.

If the last step is residuation applied to p, then p must have the form $(a \oslash b) \oslash p'$ or $p_1 \otimes p_2$. Then the premise is $p' \to (a \oslash b) \oplus q$ in the first case and $p_1 \to q/p_2$ or $p_2 \to p_1 \backslash q$ in the second case, all of which are again in the class. Furthermore if the last step is residuation applied to q, then q must have one of the forms $q' \div p'$, $c_1 \div c_2$ or $(a \oslash b) \oplus q'$ with $a \in T_2 \cup \{d\}$, $b \in T_2$, $p' \in \mathcal{P}$, $q' \in \mathcal{Q}$ and $c_1, c_2 \in \mathcal{F}$. If $q = q' \div p'$ we have as premise $p \otimes p' \to q'$ or $p' \otimes p \to q'$, both of which are again in the class. If $q = c_1 \div c_2$ we have as premise $p \otimes c_2 \to c_1$ or $c_2 \otimes p \to c_1$, which are again in the class as well. Finally if $q = (a \oslash b) \oplus q'$, the premise is either $p \oslash q' \to a \oslash b$ or $(a \oslash b) \oslash p \to q'$, of which the latter is again in the class as well. The sequent $p \oslash q' \to a \oslash b$ must have been derived by monotonicity, so one of the premises is $b \to q'$. We can easily check that the count of T_2 in q' is less than or equal to 0, while the count of T_2 in b is 1. Therefore $b \to q'$ is underivable by the count invariant, so we can conclude that $p \to q$ cannot be derived.

Finally deriving $p \to q$ with a Grishin postulate as last derivation step results in a premise which is again in the class. □

Lemma 8. *Consider the class of sequents $p \to r$ that are obtained by substituting one or more internal $\mathcal{P}$-formulas that are positive in p_1 or negative in r_1 in $p_1 \to r_1$ ($p_1 \in \mathcal{P}$, $r_1 \in \mathcal{R}$) by T_2-atoms. Sequents in this class are underivable.*

Proof. Again we show that all derivations of sequents in this class would be infinite. We select a type introduced by the substitution and call it e. First note that because atomic sequents do not have internal subformulas, there are no axioms with atomic types in this class. Now we show that we can only prove a sequent $p \rightarrow r$ from this class by proving another sequent that is also in this class. From this it follows that all derivations of sequents in this class would be infinite. Now we proceed by case distinction on the possible proof steps. Just like in Lemma 7, there are two cases that do not lead to another sequent in this class straightforwardly.

The first nontrivial case is $p \rightarrow (a \oslash b) \oplus r'$ with $a \in T_2 \cup \{d\}$, $b \in T_2$, $p \in \mathcal{P}$ and $r' \in \mathcal{R}$ preceded by right residuation, giving $p \oslash r' \rightarrow a \oslash b$, that must be preceded by monotonicity. Then the premises are $p \rightarrow a$ and $b \rightarrow r'$. If e occurs in r, then e occurs in r' as well. Then e (and b) are also internal in $b \rightarrow r'$, so this is in the class again. If not, then e must occur in p. By definition of *internal*, either a $\div$ occurs in r and therefore also in r', or e occurs within the scope of $\otimes$. In the first case, b in $b \rightarrow r'$ is internal so this sequent is again in the class. In the second case, e is internal in $p \rightarrow a$, so now this sequent is in the class again.

The second case that cannot be handled straightforwardly is monotonicity on $c_1 \div c_2 \rightarrow r' \div p'$ with $c_1, c_2 \in \mathcal{F}$, $p' \in \mathcal{P}$ and $r' \in \mathcal{R}$. Therefore the premises are $c_1 \rightarrow r'$ and $p' \rightarrow c_2$. The types c_1 and c_2 cannot contain a $\mathcal{P}$-subtype, because they are themselves $\mathcal{F}$-subtypes, so e occurs in $r' \div p'$. If e occurs in r', then $c_1 \rightarrow r'$ is in the class again. Finally if e occurs in p', then the count of T_2 in p' is at least 1, while the count of T_2 in c_2 is 0, so $p' \rightarrow c_2$ is underivable by the count invariant. □

Lemma 9. *If $p \in \mathcal{P}$ and $r \in \mathcal{R}$, then if $p \rightarrow r$ can be derived, there is a derivable sequent $p_1 \rightarrow r_1$ with $p_1 \in \mathcal{P}$ and $r_1 \in \mathcal{R}$ such that $p_1 \rightarrow r_1$ does not contain internal Grishin connectives and such that from $p_1 \rightarrow r_1$ we can derive $p \rightarrow r$ by just Grishin interactions and residuation, while every intermediate sequent in this derivation has the form $p_2 \rightarrow r_2$ with $p_2 \in \mathcal{P}$ and $r_2 \in \mathcal{R}$.*

Informally, this lemma tells us that every formula of the form $p \rightarrow r$ with $p \in \mathcal{P}$ and $r \in \mathcal{R}$ can be derived with a proof in which the last proof steps consist of making the internal Grishin connectives in $p \rightarrow r$ non-internal by Grishin interactions (and residuation). We can see this as a generalization of reversibility of steps (6)–(7) in Fig. 1.

Proof. If $p \rightarrow r$ does not contain internal Grishin connectives, we may set $p_1 = p$ and $r_1 = r$ so we are done. Now we assume $p \rightarrow r$ contains internal Grishin connectives. We inspect the proofs of sequents $p \rightarrow r$ with $p \in \mathcal{P}$ and $r \in \mathcal{R}$ (in the decidable axiomatization) and apply induction on the length of the proofs.

In the base case, where we consider all proofs of length 1, both p and r are atomic, so this sequent cannot contain internal Grishin connectives, which is a contradiction.

Now we continue with the induction step. In case the last proof step is monotonicity, the outer connective must be $\div$, so $p = a_1 \div a_2$ with $a_1, a_2 \in \mathcal{F}$ and $r = r' \div p'$ with $p' \in \mathcal{P}$ and $r' \in \mathcal{R}$, so the conclusion has the form $a_1 \div a_2 \rightarrow r' \div p'$. Then the premises are $a_1 \rightarrow r'$ and $p' \rightarrow a_2$. In case p' contains Grishin connectives, $p' \rightarrow a_2$ is underivable by Lemma 7, and in case p' does not contain Grishin connectives, r' must contain internal Grishin connectives, because $r' \div p'$

must contain internal Grishin connectives. By induction hypothesis, if $a_1 \to r'$ is derivable, it can be derived by making all internal Grishin connectives in r' non-internal by Grishin interactions and residuation. But then we could also shift the application of the monotonicity rule up in the proof, by (from bottom to top) making the Grishin connectives in r' in $a_1 \div a_2 \to r' \div p'$ non-internal by these steps (moving p' to the left by residuation, moving the Grishin connectives to the outside and moving p' back), and derive the premise thus obtained by the monotonicity rule.

If the last step in the proof is residuation applied to p, then p must have the form $(a \oslash b) \otimes p'$ or $p_1 \otimes p_2$, so we have as premise $p' \to (a \oslash b) \oplus r$ in the first case or $p_1 \to r/p_2$ or $p_2 \to p_1 \backslash r$ in the second case, on both of which the induction hypothesis applies. If the last step is residuation applied to r, then r has the form $(a \oslash b) \oplus r'$ or $r' \div p'$. In the latter case we need to prove that $p \otimes p' \to r'$ or $p' \otimes p \to r'$, so the induction hypothesis applies. In the case $r = (a \oslash b) \oplus r'$, we need to show that either $p \oslash r' \to a \oslash b$ or $(a \oslash b) \otimes p \to r'$. The induction hypothesis applies to the last sequent. The sequent $p \oslash r' \to a \oslash b$ can only have been derived by monotonicity, so we need to derive $p \to a$ and $b \to r'$. If r' contains $\div$, then b is internal, so $b \to r'$ is underivable by Lemma 8. If not, because $p \to r$ contains an internal Grishin connective and r contains no $\div$, clause (3) of the definition of *internal* applies. Therefore there is a Grishin connective within the scope of $\otimes$ in p, so there is a Grishin connective in $p \to a$. Then by induction hypothesis $p \to a$ is only derivable by first making all internal Grishin connectives in p noninternal. But then we can first mafity step.

Finally, if the last step is a Grishin interaction, we obtain a new sequent of the form $p' \to r'$ with $p' \in \mathcal{P}$ and $r' \in \mathcal{R}$, so the induction hypothesis applies.   □

Continuation of the proof of Lemma 5. First note that $d_1 \otimes \ldots \otimes d_n \to d$ (with a binary bracketing imposed on the left-hand side) is an initial sequent. We know by Lemma 6 that initial sequents have the form $p \to r$ with $p \in \mathcal{P}$ and $r \in \mathcal{R}$, and by Lemma 9 that if a sequent of this form can be derived, it can be done so by making all internal Grishin connectives non-internal by Grishin interactions (and residuation). Then we obtain a sequent of the form (6) in Fig. 1, so (5) is also derivable (by residuation). The derivation of (5) consists of some steps that reduce the right-hand side to b_n, and some steps reducing the left-hand side (necessarily also to b_n). The steps reducing the left-hand side to b_n permute with the steps reducing the right-hand side. If we reduce the right-hand side first, we obtain $a_1 \otimes \ldots \otimes a_n \to b_n$, and therefore $t_1 \ldots t_n$ is recognizable by $\mathcal{G}_1$. If we reduce the left-hand side first, we obtain (4), showing that step (4)–(5) is reversible. Furthermore, step (3)–(4) is reversible, because eliminating both b_0-atoms by monotonicity is the only way to derive (4), and the remaining residuation steps are reversible as well. Steps (1)–(3) are also reversible, so sequents of the form (1) are derivable, which shows that $t_1 \ldots t_n$ is in the permutation closure of $\mathcal{G}_2$.   □

Example 4. Consider the **AB$_\mathbf{s}$**-language in Example 3. We translate this language into an **LG**-language by the procedure of Lemma 5 such that the **LG**-language recognizes the permutation closure of the **AB$_\mathbf{s}$**-language. We keep the symbols Σ, and choose a fresh goal type s. Furthermore, we find the following type dictionary φ', with $S \in \{r, r\backslash r, r\backslash q_0, r\backslash q_3\}$.

p	$\varphi'(p)$
a	$(q_0 \oslash s) \otimes S, (q_0 \oslash q_0) \otimes S, (q_4 \oslash q_1) \otimes S, (q_4 \oslash q_2) \otimes S, (q_4 \oslash q_3) \otimes S, (q_4 \oslash q_4) \otimes S$
b	$(q_1 \oslash s) \otimes S, (q_1 \oslash q_0) \otimes S, (q_4 \oslash q_1) \otimes S, (q_4 \oslash q_2) \otimes S, (q_1 \oslash q_3) \otimes S, (q_4 \oslash q_4) \otimes S$
c	$(q_4 \oslash s) \otimes S, (q_4 \oslash q_0) \otimes S, (q_2 \oslash q_1) \otimes S, (q_4 \oslash q_2) \otimes S, (q_4 \oslash q_3) \otimes S, (q_4 \oslash q_4) \otimes S$
d	$(q_4 \oslash s) \otimes S, (q_4 \oslash q_0) \otimes S, (q_4 \oslash q_1) \otimes S, (q_3 \oslash q_2) \otimes S, (q_4 \oslash q_3) \otimes S, (q_4 \oslash q_4) \otimes S$

With those types, we can parse any permutation of strings recognized by the DFA in Example 2.

Proof (Proof of Theorem 1). As has been shown by Buszkowski [3], for every context-free language there is an equivalent product-free **NL**-grammar. Furthermore for every regular language we can find an equivalent **AB$_s$**-language by Lemma 4. It has been proved [14] that the permutation closure of the regular languages is exactly the same as the permutation closure of the context-free languages. By Lemma 5, for any product-free **NL**-language $\mathcal{L}_1$ and **AB$_s$**-language $\mathcal{L}_2$ there is an **LG**-grammar that recognizes exactly the intersection of $\mathcal{L}_1$ and the permutation closure of $\mathcal{L}_2$. Therefore we can conclude that for any language that is the intersection of a context-free language and the permutation closure of a context-free language, we can find an equivalent **LG**-grammar. $\square$

4   Examples

Corollary 1. *The language* $\{a_1{}^n \ldots a_m{}^n \mid n \in \mathbb{N}\}$ *can be recognized in* **LG** *for any* m.

Note that those languages cannot be generated by tree adjoining grammars for $m > 4$ (e.g. [8]). Therefore this example proves that **LG** generates more than tree adjoining grammars.

Proof. This language is the intersection of the permutation of the context-free language $\{(a_1 \ldots a_m)^n \mid n \in \mathbb{N}\}$ and the context-free language $\{a_1{}^* \ldots a_m{}^*\}$. $\square$

Example 5. The language $\{a^n b^n c^n d^n e^n \mid n \in \mathbb{N}\}$ can be recognized with the **LG**-grammar $\langle \Sigma, s, \varphi \rangle$ with $\Sigma = \{a, b, c, d, e\}$, goal type d and φ as follows, where $a'_1 \in \{a_1, d\}$ and $r'_5 \in \{r_5, d\}$:

p	$\varphi(p)$	
a	$(a_2 \oslash a_1) \otimes r_1$	$(a_2 \oslash a_1) \otimes (r_1 \backslash r_1)$
	$(a_2 \oslash s) \otimes r_1$	$(a_2 \oslash s) \otimes (r_1 \backslash r_1)$
b	$(a_3 \oslash a_2) \otimes (r_1 \backslash r_2)$	$(a_3 \oslash a_2) \otimes (r_2 \backslash r_2)$
c	$(a_4 \oslash a_3) \otimes (r_2 \backslash r_3)$	$(a_4 \oslash a_3) \otimes (r_3 \backslash r_3)$
d	$(a_5 \oslash a_4) \otimes (r_3 \backslash r_4)$	$(a_5 \oslash a_4) \otimes (r_4 \backslash r_4)$
e	$(a'_1 \oslash a_5) \otimes (r_4 \backslash r'_5)$	$(a'_1 \oslash a_5) \otimes (r_5 \backslash r'_5)$

Corollary 2. *A MIX-language, i.e. the permutation closure of a language consisting of strings of the form* $a_1{}^n \ldots a_m{}^n$ *with* $n \in \mathbb{N}$, *can be recognized in* **LG** *for any* m.

Proof. The language to be generated is the intersection of the permutation of the context-free language $\{(a_1 \ldots a_m)^n \mid n \in \mathbb{N}\}$ and the context-free language $\{a_k \mid 1 \leq k \leq m\}^*$.   □

Example 6. The permutation of the language $\{a^n b^n c^n \mid n \in \mathbb{N}\}$ can be recognized with the **LG**-grammar $\langle \Sigma, s, \varphi \rangle$ with $\Sigma = \{a, b, c\}$, goal type d and φ as follows, where $a' \in \{a, d\}$ and $r' \in \{r, d\}$:

p	$\varphi(\mathsf{p})$			
a	$(b \oslash a) \otimes r'$	$(b \oslash a) \otimes (r \backslash r')$	$(b \oslash s) \otimes r'$	$(b \oslash s) \otimes (r \backslash r')$
b	$(c \oslash b) \otimes r'$	$(c \oslash b) \otimes (r \backslash r')$		
c	$(a' \oslash c) \otimes r'$	$(a \oslash c) \otimes (r \backslash r')$		

5   Conclusion

We proved a new lower bound on the generative capacity of **LG**. This calculus recognizes all languages that are the intersection of a context-free language and the permutation closure of a context-free language. Therefore **LG** recognizes languages like the permutation closure of $\{a^n b^n c^n \mid n \in \mathbb{N}\}$ and for any m the language $\{a_1{}^n \ldots a_m{}^n \mid n \in \mathbb{N}\}$. The latter language cannot be generated by tree adjoining grammars, which means that this paper proves a new lower bound on the generative capacity of the Lambek–Grishin calculus. In [13] it is claimed that **LG** is equivalent to the class of languages generated by the tree adjoining grammars, but as the current paper shows, this does not apply to the general case.

There is still much unknown about the generative capacity of **LG**. For example, it is not even known whether all recognizable languages are context-sensitive. Also it is neither obvious whether **LG** can be embedded in the indexed languages, nor the other way around. Another important unsolved question is the computational complexity of parsing. Both for cognitive plausibility and practical applicability a polynomial parsing algorithm would be desirable. There exist formalisms with a polynomial parsing algorithm that possibly have a similar generative capacity to **LG**, i.e. inclusion in either direction is open, such as Range Concatenation Grammars [2] and Global Index Grammars [4].

It is clear that the *strong* generative capacity of the (binary) Lambek–Grishin calculus, i.e. the set of recognized *structures*, is less than the one of tree adjoining grammars. This is partially because tree adjoining grammars allow for branching of arbitrary order, while binary Lambek–Grishin calculus forces us to impose a binary structure on languages to be parsed. This problem disappears when we use Lambek–Grishin calculus for n-ary connectives [11] instead of binary Lambek–Grishin calculus. It remains an open question whether the n-ary Lambek–Grishin calculus can indeed recognize all structures recognized by tree adjoining grammars.

Acknowledgements

The author would like to thank Michael Moortgat and Vincent van Oostrom for their useful help and comments while supervising the MSc thesis of which this article is a result.

References

1. Bar-Hillel, Y.: A quasi-arithmetical notation for syntactic description. Language 29, 47–58 (1953)
2. Boullier, P.: Range concatenation grammars. In: New Developments in Parsing Technology, pp. 269–289. Kluwer Academic Publishers, Norwell (2004)
3. Buszkowski, W.: Generative capacity of nonassociative Lambek calculus. Bulletin of Polish Academy of Sciences: Mathematics 34, 507–516 (1986)
4. Castaño, J.M.: Global index grammars and descriptive power. J. of Logic, Lang. and Inf. 13(4), 403–419 (2004)
5. Grishin, V.N.: On a generalization of the Ajdukiewicz-Lambek system. In: Mikhailov, A.I. (ed.) Studies in Non-classical Logics and Formal Systems, pp. 315–343. Nauka, Moscow (1983)
6. Huybregts, R.: The weak inadequacy of context-free phrase structure grammars. In: Trommelen, M., de Haan, G.J., Zonneveld, W. (eds.) Van Periferie Naar Kern, pp. 81–99. Foris Publications, Dordrecht (1984)
7. Joshi, A., Levy, L.S., Takahashi, M.: Tree adjunct grammars. Journal of Computer and System Sciences 10(1), 136–163 (1975)
8. Joshi, A., Schabes, Y.: Tree-adjoining grammars. In: Handbook of Formal Languages. Beyond Words, vol. 3, pp. 69–123. Springer, Inc., New York (1997)
9. Kandulski, M.: The equivalence of nonassociative Lambek categorial grammars and context-free grammars. Zeitschrift für Mathematische Logik und Grundlagen der Mathematik 34(1), 41–52 (1988)
10. Lambek, J.: On the calculus of syntactic types. In: Jacobsen, R. (ed.) Structure of Language and its Mathematical Aspects. Proceedings of Symposia in Applied Mathematics, vol. XII, pp. 166–178. American Mathematical Society, Providence (1961)
11. Melissen, M.: Lambek–Grishin calculus extended to connectives of arbitrary arity. In: Proceedings of the 20th Belgian-Netherlands Conference on Artificial Intelligence (2008)
12. Moortgat, M.: Symmetries in natural language syntax and semantics: The lambek-grishin calculus. In: Leivant, D., de Queiroz, R. (eds.) WoLLIC 2007. LNCS, vol. 4576, pp. 264–284. Springer, Heidelberg (2007)
13. Moot, R.: Lambek grammars, tree adjoining grammars and hyperedge replacement grammars. In: Proceedings of the TAG+ Conference, HAL - CCSD (2008)
14. Parikh, R.: On context-free languages. J. ACM 13(4), 570–581 (1966)
15. Pentus, M.: Lambek grammars are context free. In: Proceedings of the 8th Annual IEEE Symposium on Logic in Computer Science, pp. 429–433. IEEE Computer Society Press, Los Alamitos (1993)
16. Sipser, M.: Introduction to the Theory of Computation. Course Technology (December 1996)
17. Van Benthem, J.: Language in action: categories, lambdas and dynamic logic. Studies in logic and the foundations of mathematics, vol. 130. North-Holland, Amsterdam (1991)

A Savateev-Style Parsing Algorithm for Pregroup Grammars

Katarzyna Moroz

Faculty of Mathematics and Computer Science,
Adam Mickiewicz University, Poznań
moroz@amu.edu.pl

Abstract. We present a new cubic parsing algorithm for ambiguous pregroup grammars. It modifies the recognition algorithm of Savateev [10] for categorial grammars based on $\mathbf{L}^{\backslash}$. We show the correctness of the algorithm and give some examples. We compare our algorithm with the algorithm of Oehrle [8] for pregroup grammars and the algorithm of Savateev [10].

1   Introduction

Pregroups were introduced by Lambek [5] as an algebraic tool for the syntactical analysis of sentences. The idea is that words have syntactical properties that can be described by a finite set of pregroup types. The problem of parsing pregroup grammars has been a subject of study in recent years (see the references).

Oehrle [8] proposed a recognising algorithm based on graph theoretic representations of lexical types and their complexes. The algorithm can be amended to become a parsing algorithm. Béchet [1] presents a recognition algorithm based on a restricted form of partial composition called functional composition. However, the algorithm employs a construction of a context-free grammar whose size is exponential with respect to the given pregroup grammar. A polynomial construction is given in [4]. Another approach to parsing pregroup grammars was proposed by Preller [9]. It yields a linear algorithm for unambiguous pregroup grammars with some restrictions imposed on the type lexicon.

Savateev [10] provides a characterisation of derivability in $\mathbf{L}^{\backslash}$ (the Lambek calculus with one residual \ and without product), which is, essentially, a translation of $\mathbf{L}^{\backslash}$ in the right pregroup calculus with some additional constraint. Then, he presents a recognition algorithm for $\mathbf{L}^{\backslash}$-grammars, which uses this translation. Our algorithm modifies the latter. There are essential differences: (1) we regard poset rules and both left and right adjoints, while Savateev considers right adjoints only and no poset rules, (2) we drop the extra constraint needed for the translation of $\mathbf{L}^{\backslash}$. (3) Savateev's algorithm depends on some special properties of $\mathbf{L}^{\backslash}$-types, which is not the case for our algorithm. (Actually, Savateev does not explicitly refer to pregroups, but his translation can be interpreted in terms of pregroups.) Further we present briefly the algorithm of Savateev. We provide a complete proof of the correctness of our algorithm. At the end, we show how the

P. de Groote, M. Egg, and L. Kallmeyer (Eds.): FG 2009, LNAI 5591, pp. 133–149, 2011.

algorithm can be extended to return the reduction, thus becoming a full parsing algorithm. We also present some examples.

2   Preliminaries

A *pregroup* is a structure $M = (M, \leqslant, \cdot, l, r, 1)$ such that $(M, \leqslant, \cdot, 1)$ is a partially ordered monoid, and l, r are unary operations on M, satisfying the adjoint laws:

$$(Al)\ a^l a \leqslant 1 \leqslant aa^l, \quad (Ar)\ aa^r \leqslant 1 \leqslant a^r a,$$

for all $a \in M$. Pregroups are a generalization of partially ordered groups. a^l (resp. a^r) is called the *left* (resp. *right*) *adjoint* of a.

A *right pregroup* is a partially ordered monoid in which for each element a there exists a *right adjoint* a^r satisfying the law (Ar).

It is worth noticing that the following laws are easily derivable from (Al), (Ar):

$$1^l = 1 = 1^r,\ a^{lr} = a = a^{rl},\ (ab)^l = b^l a^l,\ (ab)^r = b^r a^r,$$

$$a \leqslant b \text{ iff } b^l \leqslant a^l \text{ iff } b^r \leqslant a^r.$$

One defines *iterated adjoints*: $a^{(n)} = a^{rr...r}$ and $a^{(-n)} = a^{ll...l}$, where n is a non-negative integer and a any element of **M**. The following laws are provable:

$$(a^{(n)})^l = a^{(n-1)},$$
$$(a^{(n)})^r = a^{(n+1)},$$
$$(a_1^{(n_1)}...a_k^{(n_k)})^r = a_k^{(n_k+1)}...a_1^{(n_1+1)}$$
$$(a_1^{(n_1)}...a_k^{(n_k)})^l = a_k^{(n_k-1)}...a_1^{(n_1-1)}$$
$$a^{(n)}a^{(n+1)} \leqslant 1 \leqslant a^{(n+1)}a^{(n)}, \text{ for any integer } n.$$

Lambek's approach to syntactic analysis is based on the notion of a *free pregroup*, generated by a finite poset $(P, \leqslant)$.

Elements of P are called *atoms* and denoted by letters p, q, r, s.

Expressions of the form $p^{(n)}$, for any $p \in P$ and any integer n are *simple terms*.

A *term* is a finite sequence (string) of simple terms. Terms are also called *types*. They are denoted by $\mathbb{A}, \mathbb{B}, \mathbb{C}$. If $p \leqslant q$ in P, then we write $p^{(n)} \leqslant q^{(n)}$, if n is even, and $q^{(n)} \leqslant p^{(n)}$, if n is odd.

One defines a binary relation $\Rightarrow$ on the set of terms as the least reflexive and transitive relation, satisfying the clauses:

(CON) $\mathbb{A}, p^{(n)}, p^{(n+1)}, \mathbb{B} \Rightarrow \mathbb{A}, \mathbb{B}$,
(EXP) $\mathbb{A}, \mathbb{B} \Rightarrow \mathbb{A}, p^{(n+1)}, p^{(n)}, \mathbb{B}$,
(IND) $\mathbb{A}, p^{(n)}, \mathbb{B} \Rightarrow \mathbb{A}, q^{(n)}, \mathbb{B}$, if $p^{(n)} \leqslant q^{(n)}$.

(CON), (EXP), (IND) are called Contraction, Expansion and Induced Step, respectively. They can be treated as rules of a term rewriting system. $\mathbb{A} \Rightarrow \mathbb{B}$ is true iff $\mathbb{A}$ can be rewritten into $\mathbb{B}$ by a finite number of applications of these rules. This rewriting system is Lambek's original form of the logic of pregroups. This logic is also called *Compact Bilinear Logic* (**CBL**).

One can also define a *Generalized Contraction* that combines Contraction with Inducesd Step:

(GCON) $\mathbb{A}p^{(n)}q^{(n+1)}\mathbb{B} \Rightarrow \mathbb{AB}$ if $p \leqslant q$ and n is even, or $q \leqslant p$ and n is odd.

Lambek [5] shows an important property called *the switching lemma*:

Lemma 1. *If* $\mathbb{A} \Rightarrow \mathbb{B}$ *then there exists* $\mathbb{C}$ *such that* $\mathbb{A} \Rightarrow \mathbb{C}$ *without (EXP) and* $\mathbb{C} \Rightarrow \mathbb{B}$ *without (CON).*

One can easily see as a consequence of this theorem that if $\mathbb{A} \Rightarrow t$, where t is a simple term or ε, then $\mathbb{A}$ can be reduced to t by (CON) and (IND) only. Such reductions are easily computable and can be simulated by a context-free grammar. This yields the polynomial time decidability of **CBL** [2,3].

A *pregroup grammar* is a quintuple $G = (\Sigma, P, \leqslant, s, I)$ such that Σ is a nonempty, finite alphabet, $(P, \leqslant)$ is a finite poset, $s \in P$, and I is a finite relation between symbols from Σ and nonempty types (on P). For $a \in \Sigma$, $I(a)$ denotes the set of all types $\mathbb{A}$ such that $(a, \mathbb{A}) \in I$. Let $x \in \Sigma^+$, $x = a_1...a_n$ $(a_i \in \Sigma)$. One says that the grammar G assigns type $\mathbb{B}$ to x, if there exist types $\mathbb{A}_i \in I(a_i)$, $i = 1, ..., n$, such that $\mathbb{A}_1, ..., \mathbb{A}_n \Rightarrow \mathbb{B}$ in **CBL**; we write $x \to_G \mathbb{B}$. The set $L(G) = \{x \in \Sigma^+ : x \to_G s\}$ is called the language of G. We should notice that the condition $\mathbb{A}_{\Join}...\mathbb{A}_{\ltimes} \Rightarrow \mathbb{B}$ in **CBL** is equivalent to the condition $\mathbb{A}_{\Join}...\mathbb{A}_{\ltimes}\mathbb{B}^{(1)} \Rightarrow \varepsilon$.

It is useful to present reductions by a set of *links*. A link indicates a pair of simple terms that reduce to 1. If a whole string reduces to 1 then any simple term appearing in the string is connected by a link with another simple term. Each term can be a member of only one link and links cannot cross. We can see the links in the following example [6]:

$$
\begin{array}{cccc}
I & will & meet & him. \\
\pi_1) \;(\;\pi^r \; s_1 \; j^l) & (i \; o^l) & (o) & s^r
\end{array}
$$

where there are the following basic types:

π_1 - first person singular subject
π - subject
i - infinitive of intransitive verb
j - infinitive of any complete verb phrase
o - direct object
s_1 - declarative sentence in present tense
s - sentence

and the following partial order:

$$\pi_1 \leqslant \pi, \quad i \leqslant j, \quad s_1 \leqslant s$$

3   Parsing Algorithm

Our goal is to check whether a given string x is a member of the language generated by the pregroup grammar G: $x \in L(G)$. If this is the case, we want to obtain the appropriate derivation.

We define the algorithm in a style proposed by Savateev [10] for unidirectional Lambek calculus. It is a dynamic algorithm working on a special form of a string containing all possible type assignments for words of the sentence to parse.

3.1   Definition of the Function M

We fix $G = (\Sigma, P, \leqslant, s, I)$ and $x \in \Sigma^+$, $x = a_1...a_n$. We use special symbols $*, \langle, \rangle$. In what follows, we write 1 for ε. Let us denote:
- $\mathcal{Z}$ - the set of integers
- $T = \{p^{(n)} : p \in P, n \in \mathcal{Z}\}$ - the set of simple terms
- $\mathbb{A}, \mathbb{B}, \mathbb{C}, \mathbb{D}, ..$ - elements of T^*
- $k^a = |I(a)|$
- $\mathbb{A}_j^a$ - the j-th possible assignment of type to a, $1 \leqslant j \leqslant k^a$ (hence $I(a) = \{\mathbb{A}_1^a, ...\mathbb{A}_{k^a}^a\}$)
- $Q^a = \langle *\mathbb{A}_1^a * \mathbb{A}_2^a * ... * \mathbb{A}_{k^a}^a *\rangle$
- $W^x \in (T \cup \{*, \langle, \rangle\})^* = Q^{a_1}...Q^{a_n}\langle *s^{(1)}$
- $W_i^x, 1 \leqslant i \leqslant |W^x|$ - the i-th symbol of the string W^x
- $W_{[i,j]}^x, 1 \leqslant i < j \leqslant |W^x|$ - the substring of W^x: $W_i^x W_{i+1}^x...W_j^x$

Let $M(i,j), 1 \leqslant i < j \leqslant |W^x|$ be a function such that $M(i,j) = 1$ iff one of the following conditions holds:

- **M1.** $W_{[i,j]}^x \in T^+$ and it reduces to 1.
- **M2a.** $W_{[i,j]}^x$ is of the form $\langle ... \rangle...\langle V * \mathbb{C}$, where:
  - $\mathbb{C} \in T^+$
  - V contains no angle brackets
  - in $W_{[i,j]}^x$ there are g ($g \geqslant 0$) pairs of matched angle brackets; for the h-th pair of them there is a substring of the form $*\mathbb{D}_h*$ in between them such that $\mathbb{D}_h \in T^+$ and the string $\mathbb{D}_1...\mathbb{D}_g\mathbb{C}$ reduces to 1
- **M2b.** $W_{[i,j]}^x$ is of the form $\mathbb{C} * U\rangle...\langle ... \rangle$, where:
  - $\mathbb{C} \in T^+$
  - U contains no angle brackets
  - in $W_{[i,j]}^x$ there are g ($g \geqslant 0$) pairs of matched angle brackets; for the h-th pair of them there is a substring of the form $*\mathbb{D}_h*$ in between them such that $\mathbb{D}_h \in T^+$ and the string $\mathbb{C}\mathbb{D}_1...\mathbb{D}_g$ reduces to 1
- **M3.** $W_{[i,j]}^x$ is of the form $\mathbb{D} * U\rangle...\langle V * \mathbb{C}$, where:
  - $\mathbb{C}, \mathbb{D} \in T^+$
  - U,V contains no angle brackets
  - in $W_{[i,j]}^x$ there are g ($g \geqslant 0$) pairs of matched angle brackets; for the h-th pair of them there is a substring of the form $*\mathbb{E}_h*$ in between them such that $\mathbb{E}_h \in T^+$ and the string $\mathbb{D}\mathbb{E}_1...\mathbb{E}_g\mathbb{C}$ reduces to 1

- **M4.** $W_{[i,j]}^x$ is of the form $\langle...\rangle...\langle...\rangle$, where:

  - in $W_{[i,j]}^x$ there are g $(g \geqslant 1)$ pairs of matched angle brackets; for the h-th pair of them there is a substring of the form $*\mathbb{E}_h*$ in between them such that $\mathbb{E}_h \in T^+$ and the string $\mathbb{E}_1...\mathbb{E}_g$ reduces to 1

In all other cases $M(i, j) = 0$.

3.2   The Recognition Algorithm

We compute $M(i, j)$ dynamically. $M(i, i + 1) = 1$ only if $W_i^x = p^{(m)}$, $W_{i+1}^x = q^{(m+1)}$ and the following condition holds

(GC) $p, q \in P, m \in \mathcal{Z}$ and $p \leqslant q$ in P if m is even or $q \leqslant p$ in P if m is odd.

The other initial case is when $W_{[i,j]}^x$ is of the form $p^{(m)} * U\rangle\langle V * q^{(m+1)}$, the condition (GC) holds and strings U, V contain no angle brackets. Then we put $M(i, j) = 1$.

When we already know $M(g, h)$, $1 \leqslant g < h \leqslant |W^x|$ and $h - g < j - i$, we can compute $M(i, j)$. There are several cases:

- **A1a.** $W_i^x, W_j^x \in T$. If there exists k such that $i < k < j - 1$ and $W_k^x \in T$, $W_{(k+1)}^x \in T$ and $M(i, k) = 1$ and $M(k + 1, j) = 1$, then we put $M(i, j) = 1$.
- **A1b.** $W_i^x, W_j^x \in T$. If there exists k such that $i < k < j - 1$ and $W_k^x = \rangle$, $W_{(k+1)}^x = \langle$ and $M(i, k) = 1$ and $M(k + 1, j) = 1$, then we put $M(i, j) = 1$.
- **A2.** $W_i^x = p^{(m)}$, $W_j^x = q^{(m+1)}$ and the condition (GC) holds. If $M(i + 1, j - 1) = 1$, then $M(i, j) = 1$.
- **A3a.** $W_{[i,j]}^x$ is of the form $\langle...\rangle...\langle...p^{(m)}$, $p \in P, m \in \mathcal{Z}$. If there exists k such that $i < k < j$ and $W_k^x = *$, $W_{[i+1,k]}^x$ contains no angle brackets and $M(k + 1, j) = 1$ then $M(i, j) = 1$
- **A3b.** $W_{[i,j]}^x$ is of the form $p^{(m)}...\rangle...\langle...\rangle$, $p \in P, m \in \mathcal{Z}$. If there exists k such that $i < k < j$ and $W_k^x = *$, $W_{[k,j-1]}^x$ contains no angle brackets and $M(i, k - 1) = 1$. Then we put $M(i, j) = 1$
- **A4a.** $W_{[i,j]}^x$ is of the form $p^{(m)}*...\rangle...\langle...q^{(m+1)}$ and the condition (GC) holds. If $M(k, j - 1) = 1$, where k is the position of the first left angle bracket in the string $W_{[i,j]}^x$. Then we put $M(i, j) = 1$
- **A4b.** $W_{[i,j]}^x$ is of the form $p^{(m)}...\rangle...\langle...*q^{(m+1)}$ and the condition (GC) holds. If $M(i + 1, k) = 1$, where k is the position of the last right angle bracket in the string $W_{[i,j]}^x$ then $M(i, j) = 1$
- **A4c.** $W_{[i,j]}^x$ is of the form $p^{(m)} * ...\rangle...\langle... * q^{(m+1)}$ where the string "..." in between the angle brackets is not empty and the condition (GC) holds. If $M(k, k') = 1$, where k is the position of the first left angle bracket in the string $W_{[i,j]}^x$ and k' is the position of the last right angle bracket in the string $W_{[i,j]}^x$ then $M(i, j) = 1$

- **A5.** $W_{[i,j]}^x$ is of the form $\langle...\rangle...\langle...\rangle$. If $M(k,k') = 1$, where W_k^x is a simple term in between the first pair of angle brackets, $W_{k'}^x$ is a simple term in between last pair of angle brackets in the string $W_{[i,j]}^x$ and $W_{k-1}^x = *$ and $W_{k'+1}^x = *$, then $M(i,j) = 1$.

In all other cases $M(i,j) = 0$.

Theorem 1. *The algorithm computes $M(i,j)$ correctly.*

Proof. We will show at first that if the algorithm computes $M(i,j) = 1$, then $M(i,j) = 1$ according to the definition of **M**. We will prove it by induction on the length of the string.

For strings of length 2, the algorithm computes $M(i,j) = 1$ only in case when $W_i^x = p^{(m)}$ and $W_{i+1}^x = q^{(m+1)}$ and the condition (GC) holds. $W_{[i,j]}^x$ is then of the form **(M1)** since $W_{[i,j]}^x \in T^+$ and the string $W_{[i,j]}^x$ reduces to 1. So $M(i,j) = 1$ according to the definition of **M**.

The other initial case when the algorithm computes $M(i,j) = 1$ is when $W_{[i,j]}^x$ is of the form $p^{(m)} * U\rangle\langle V * q^{(m+1)}$, the condition (GC) holds and the strings U and V contain no angle brackets. $W_{[i,j]}^x$ is then of the form **(M3)** since we can assume $\mathbb{D} = p^{(m)}$ and $\mathbb{C} = q^{(m+1)}$ and $g = 0$. So $\mathbb{DC}$ reduces to 1. Hence $M(i,j) = 1$ according to the definition of **M**.

Let us consider now the recursive cases when the algorithm computes $M(i,j) = 1$ (all cases of the description of the algorithm).

- **A1a.** $W_i^x, W_j^x \in T$ and there exists k such that $i < k < j - 1$ and $W_k^x \in T$, $W_{(k+1)}^x \in T$, $M(i,k) = 1$ and $M(k+1,j) = 1$. Then the substrings $W_{[i,k]}^x$ and $W_{[k+1,j]}^x$ are shorter than $W_{[i,j]}^x$ therefore by the induction hypothesis both $M(i,k)$ and $M(k+1,j)$ are equal to 1 according to the definition of **M**. $W_{[i,k]}^x$ and $W_{[k+1,j]}^x$ can be of the form **(M1)** or **(M3)**. There are the following cases.
  - If both substrings $W_{[i,k]}^x$ and $W_{[k+1,j]}^x$ are of the form **(M1)** then $W_{[i,j]}^x$ also consists of simple terms and reduces to 1, as both $W_{[i,k]}^x$ and $W_{[k+1,j]}^x$ reduce to 1. $W_{[i,j]}^x$ is therefore of the form **(M1)**. Hence $M(i,j) = 1$ in accordance with the definition of **M**.
  - $W_{[i,k]}^x$ is of the form **(M1)**. Therefore it consists of simple terms and reduces to 1 and $W_{[k+1,j]}^x$ is of the form **(M3)**. Then $W_{[i,j]}^x$ is also of the form **(M3)**. Hence $M(i,j) = 1$ according to the definition of **M**.
  - $W_{[i,k]}^x$ is of the form **(M3)** and $W_{[k+1,j]}^x$ is of the form **(M1)**. So it consists of simple terms and reduces to 1. Then $W_{[i,j]}^x$ is also of the form **(M3)**. Hence $M(i,j) = 1$ according to the definition of **M**.
  - $W_{[i,k]}^x$ and $W_{[k+1,j]}^x$ are of the form **(M3)**. Hence the whole string $W_{[i,j]}^x$ is also of the form **(M3)**. Then $M(i,j) = 1$ in accordance with the definition of **M**.
- **A1b.** $W_i^x, W_j^x \in T$ and there exists k such that $i < k < j - 1$ and W_k^x is a right angle bracket, $W_{(k+1)}^x$ is a left angle bracket, $M(i,k) = 1$ and $M(k+1,j) = 1$.

Then the substrings $W_{[i,k]}^x$ and $W_{[k+1,j]}^x$ are shorter than $W_{[i,j]}^x$ therefore by the induction hypothesis both $M(i,k)$ and $M(k+1,j)$ are equal to 1 according to the definition of $\mathbf{M}$. $W_{[i,k]}^x$ is of the form **(M2b)** so there is a string $\mathbb{C}\mathbb{D}_1...\mathbb{D}_{g_1}$ that reduces to 1 (such that the first simple term of $\mathbb{C}$ is W_i^x, $\mathbb{D}_h \in T^+$ is a substring in between the h-th pair of matching angle brackets in the string $W_{[i,k]}^x$). The substring $W_{[k+1,j]}^x$ is of the form **(M2a)** so there is a string $\mathbb{E}_1...\mathbb{E}_{g_2}\mathbb{F}$ that reduces to 1 (such that the last simple term of $\mathbb{F}$ is W_j^x, $\mathbb{E}_h \in T^+$ is a substring in between the h-th pair of matching angle brackets in the string $W_{[k+1,j]}^x$). So the whole string $W_{[i,j]}^x$ is of the form **(M3)** (with the string $\mathbb{C}\mathbb{D}_1...\mathbb{D}_{g_1}\mathbb{E}_1...\mathbb{E}_{g_2}\mathbb{F}$ reducing to 1).

A2. $W_i^x = p^{(m)}$, $W_j^x = q^{(m+1)}$, the condition (GC) holds and $M(i+1,j-1) = 1$. $M(i+1,j-1) = 1$ is computed according to the definition of $\mathbf{M}$ by the induction hypothesis, as $W_{[i+1,j-1]}^x$ is shorter than $W_{[i,j]}^x$. Then the substring $W_{[i+1,j-1]}^x$ can be:

- of the form **(M1)**, then the whole string $W_{[i,j]}^x$ consists of simple terms and reduces to 1, so it is also of the form **(M1)**. Hence $M(i,j) = 1$ according to the definition of $\mathbf{M}$.

- of the form **(M3)**, that is $\mathbb{D} * U\rangle...\langle V * \mathbb{C}$, where U and V contain no angle brackets and the string $\mathbb{D}\mathbb{E}_1...\mathbb{E}_g\mathbb{C}$ reduces to 1. We can assume $\mathbb{D}' = p^{(m)}\mathbb{D}$ and $\mathbb{C}' = \mathbb{C}q^{(m+1)}$. Then $W_{[i,j]}^x = \mathbb{D}' * U\rangle...\langle V * \mathbb{C}'$ and the string $\mathbb{D}'\mathbb{E}_1...\mathbb{E}_g\mathbb{C}'$ reduces to 1. Hence $M(i,j) = 1$ according to the definition of $\mathbf{M}$.

We should notice that the string $W_{[i+1,j-1]}^x$ cannot be of the form **(M2a)**, **(M2b)** or **(M4)** since a simple term can be followed (preceded) only by another simple term or an asterisk.

A3a. $W_{[i,j]}^x$ is of the form $\langle...\rangle...\langle...p^{(m)}$, $p \in P, m \in \mathcal{Z}$ and there exists k such that $i < k < j$ and $W_k^x = *$, the string $W_{[i+1,k]}^x$ contains no angle brackets and $M(k+1,j) = 1$. $W_{[k+1,j]}^x$ is shorter than $W_{[i,j]}^x$ so $M(k+1,j) = 1$ is computed according to the definition of $\mathbf{M}$ by the induction hypothesis. The string $W_{[k+1,j]}^x$ ends with a simple term, so it can be of the form **(M1)**, **(M2a)** or **(M3)**. But it cannot begin with a left angle bracket and it contains angle brackets, so it must be of the form **(M3)**. So there is a string $\mathbb{C}\mathbb{E}_1...\mathbb{E}_g\mathbb{D}$ reducing to 1. Hence $W_{[i,j]}^x$ is of the form **(M2a)** with the string $\mathbb{C}\mathbb{E}_1...\mathbb{E}_g\mathbb{D}$ reducing to 1. Therefore $M(i,j) = 1$ according to the definition of $\mathbf{M}$.

A3b. $W_{[i,j]}^x$ is of the form $p^{(m)}...\rangle...\langle...\rangle$, where $p \in P, m \in \mathcal{Z}$ and there exists k such that $i < k < j$, $W_k^x = *$, $W_{[k,j-1]}^x$ contains no angle brackets and $M(i,k-1) = 1$. The string $W_{[i,k-1]}^x$ is shorter than $W_{[i,j]}^x$ so M(i,k-1)=1 is computed according to the definition of $\mathbf{M}$, by the induction hypothesis. The string $W_{[i,k-1]}^x$ begins with a simple term, so it can be of the form **(M1)**, **(M2b)** or **(M3)**. But it cannot end with a right angle bracket and it contains angle brackets, so it must be of the form **(M3)**. So there is a string $\mathbb{C}\mathbb{E}_1...\mathbb{E}_g\mathbb{D}$ reducing to 1. Hence $W_{[i,j]}^x$ is of the form **(M2b)** with the string $\mathbb{C}\mathbb{E}_1...\mathbb{E}_g\mathbb{D}$ reducing to 1. Therefore $M(i,j) = 1$ according to the definition of $\mathbf{M}$.

A4a. $W_{[i,j]}^x$ is of the form $p^{(m)} * ...\rangle...\langle...q^{(m+1)}$, the condition (GC) holds and $M(k, j-1) = 1$, where k is the position of the first left angle bracket in the string $W_{[i,j]}^x$. $W_{[k,j-1]}^x$ is shorter than $W_{[i,j]}^x$. Hence, by the induction hypothesis, $M(k, j-1) = 1$ according to the definition of $\mathbf{M}$. The string $W_{[k,j-1]}^x$ must be therefore of the form **(M2a)** since it is the only form starting with an angle bracket and ending with something different from a bracket. Hence $W_{[k,j-1]}^x = \langle...\rangle...\langle V * \mathbb{C}$, where $\mathbb{C}$ is the string of simple terms, V contains no angle brackets and there are g pairs of matched angle brackets, for the h-th of them there is the substring $*\mathbb{E}_h*$ in between them, such that $\mathbb{E}_h \in T^+$ and $\mathbb{E}_1...\mathbb{E}_g\mathbb{C}$ reduces to 1. Let $\mathbb{C}' = \mathbb{C}q^{(m+1)}$, $\mathbb{D} = p^{(m)}$. Then $\mathbb{D}\mathbb{E}_1...\mathbb{E}_g\mathbb{C}'$ reduces to 1, and the string $W_{[i,j]}^x$ is therefore of the form **(M3)**. Then $M(i,j) = 1$ in accordance with the definition of $\mathbf{M}$.

A4b. $W_{[i,j]}^x$ is of the form $p^{(m)}...\rangle...\langle... * q^{(m+1)}$, the condition (GC) holds, and $M(i+1, k) = 1$, where k the position of the last right angle bracket in the string $W_{[i,j]}^x$. $W_{[i+1,k]}^x$ is shorter than $W_{[i,j]}^x$. Hence, by the induction hypothesis, $M(i+1, k) = 1$ according to the definition of $\mathbf{M}$. The string $W_{[i+1,k]}^x$ must be therefore of the form **(M2b)** since it is the only form ending with an angle bracket but starting with something different from a bracket. Hence $W_{[i+1,k]}^x = \mathbb{D}*U\rangle...\langle...\rangle$, where $\mathbb{D}$ is the string of simple terms, U contains no angle brackets and there are g pairs of matched angle brackets, for the h-th of them there is the substring $*\mathbb{E}_h*$ in between them, such that $\mathbb{E}_h \in T^+$ and $\mathbb{D}\mathbb{E}_1...\mathbb{E}_g$ reduces to 1. Let $\mathbb{D}' = p^{(m)}\mathbb{D}$, $\mathbb{C} = q^{(m+1)}$. Then $\mathbb{D}'\mathbb{E}_1...\mathbb{E}_g\mathbb{C}$ reduces to 1, and the string $W_{[i,j]}^x$ is therefore of the form **(M3)**. Then $M(i,j) = 1$ in accordance with the definition of $\mathbf{M}$.

A4c. $W_{[i,j]}^x$ is of the form $p^{(m)}*...\rangle...\langle...*q^{(m+1)}$, where the string "..." in between the angle brackets is not empty and the condition (GC) holds. $M(k, k') = 1$, where k is the position of the first left angle bracket in the string $W_{[i,j]}^x$ and k' the position of the last right angle bracket in the string $W_{[i,j]}^x$. $W_{[k,k']}^x$ is shorter than $W_{[i,j]}^x$. Hence, by the induction hypothesis, $M(k, k') = 1$ according to the definition of $\mathbf{M}$. The string $W_{[k,k']}^x$ must be therefore of the form **(M4)**. Hence $W_{[k,k']}^x = \langle...\rangle...\langle...\rangle$, where there are g pairs of matched angle brackets, for the h-th of them there is the substring $*\mathbb{E}_h*$ in between them, such that $\mathbb{E}_h \in T^+$ and $\mathbb{E}_1...\mathbb{E}_g$ reduces to 1. Let $\mathbb{D} = p^{(m)}$, $\mathbb{C} = q^{(m+1)}$, $U = W_{[i+2,k-1]}^x$ (if $k > i + 3$ else $U = \varepsilon$) and $V = W_{[k'+1,j-2]}^x$ (if $k' < j - 3$ else $V = \varepsilon$). Then $\mathbb{D}\mathbb{E}_1...\mathbb{E}_g\mathbb{C}$ reduces to 1, and the string $W_{[i,j]}^x$ is therefore of the form **(M3)**. Then $M(i,j) = 1$ according to the definition of $\mathbf{M}$.

A5. $W_{[i,j]}^x$ is of the form $\langle...\rangle...\langle...\rangle$, and $M(k, k') = 1$, where W_k^x is a simple term in between the first pair of angle brackets and $W_{k'}^x$ is a simple term in between last pair of angle brackets in the string $W_{[i,j]}^x$ and $W_{k-1}^x = *$ and $W_{k'+1}^x = *$. $W_{[k,k']}^x$ is shorter than $W_{[i,j]}^x$ hence $M(k, k') = 1$ according to the definition of $\mathbf{M}$. $W_{[k,k']}^x$ must be then of the form **(M3)**, so the whole string $W_{[i,j]}^x$ is of the form **(M4)** and therefore $M(i,j) = 1$ in accordance with the definition of $\mathbf{M}$.

We will prove that the algorithm finds correctly all substrings for which the function $M(i,j) = 1$ by induction on the length of the substring. Let us notice that there are no such substrings that contain asterisk but no angle brackets.

The only strings of length 2 for which $M(i,j) = 1$ are of the form $p^{(m)}q^{(m+1)}$ where the condition (GC) holds (that is of the form **(M1)**), and the algorithm finds them correctly.

Let us consider now the substrings of the length $l > 2$ such that for all $l' < l$ the algorithm finds the substrings of the length l' correctly. (all forms of the definition of **M**)

- 1. $W^x_{[i,j]}$ is of the form **(M1)** that is $W^x_{[i,j]} \in T^+$ and it reduces to 1. There are 2 possible cases:
  - $W^x_i W^x_j$ does not reduce to 1 in the whole reduction of the string $W^x_{[i,j]}$ to 1. Then $W^x_{[i,j]}$ can be divided into 2 substrings. Each of them reduces to 1, is shorter than $W^x_{[i,j]}$, so they are found by the algorithm correctly (by the induction hypothesis). Hence by **(A1a)**, $M(i,j) = 1$.
  - $W^x_i W^x_j$ reduces to 1 in the whole reduction of $W^x_{[i,j]}$ to 1. Then $W^x_{[i+1,j-1]}$ is shorter substring reducing to 1. By the induction hypothesis the string is found by the algorithm correctly, so $M(i,j) = 1$ (case **(A2)**).
- 2a. $W^x_{[i,j]}$ is of the form **(M2a)**. If $M(i,j) = 1$ then there exists a substring of the form $*\mathbb{D}_1*$ in between the first pair of angle brackets such that $\mathbb{D}_1$ takes part in some reduction to 1 (if $g > 0$). Let i' be the position of the first simple term in $\mathbb{D}_1$. The substring $W^x_{[i',j]}$ is of the form **(M3)**, is shorter than $W^x_{[i,j]}$ and $M(i',j) = 1$. By the induction hypothesis, the string is found by the algorithm correctly, so $M(i,j) = 1$ (case **(A3a)**). If $g = 0$, then $\mathbb{C}$ reduces to 1 and it is of the form **(M1)** and shorter than $W^x_{[i,j]}$. So, by the induction hypothesis it is found by the algorithm correctly. Then $M(i,j) = 1$ by **(A3a)**.
- 2b. $W^x_{[i,j]}$ is of the form **(M2b)**. If $M(i,j) = 1$ then there exists the substring of the form $*\mathbb{D}_g*$ (if $g > 0$) in between the last pair of angle brackets such that $\mathbb{D}_g$ takes part in some reduction to 1. Let j' be the position of the last simple term in $\mathbb{D}_g$. The substring $W^x_{[i,j']}$ is of the form **(M3)**, is shorter than $W^x_{[i,j]}$ and $M(i,j') = 1$. By the induction hypothesis the string is found by the algorithm correctly, so $M(i,j) = 1$ (case **(A3b)**). If $g = 0$ then $\mathbb{C}$ reduces to 1 and it is of the form **(M1)** and shorter than $W^x_{[i,j]}$, so by the induction hypothesis it is found by the algorithm correctly. Then $M(i,j) = 1$ by **(A3b)**.
- 3. $W^x_{[i,j]}$ is of the form **(M3)**. $W^x_i W^x_j$ takes part in the reduction to 1 in $W^x_{[i,j]}$. Let us assume: $W^x_i W^x_{i'}$ reduce to 1 where $i' \neq j$. Then $W^x_{i'+1}$ can be:
  - simple term. Then $W^x_{[i,i']}$ and $W^x_{[i'+1,j]}$ are of the form **(M1)** or **(M3)**, they are shorter and both $M(i,i') = 1$ and $M(i'+1,j) = 1$. Hence, by the induction hypothesis, these strings are found by the algorithm correctly, so $M(i,j) = 1$ (case **(A1a)**).
  - asterisk. Then $W^x_{[i,i']}$ is of the form **(M1)** or **(M3)**, it is shorter and $M(i,i') = 1$. Hence, by the induction hypothesis, the string is found

by the algorithm correctly. Let k be the position of the first right angle bracket following i'. $W^x_{[i,k]}$ is shorter than $W^x_{[i,j]}$ and it is of the form **(M2b)**. So, by the induction hypothesis $M(i,k) = 1$ (case **(A3b)**). Similarly the substring $W^x_{[k+1,j]}$ is of the form **(M2a)** is shorter than $W^x_{[i,j]}$. So, by the induction hypothesis $M(k + 1, j) = 1$ (case **(A3a)**). Hence $M(i,j) = 1$ by the induction hypothesis (case **(A1b)**).

Let us assume $i' = j$ so $W^x_i W^x_j$ reduces to 1. There are the following cases:

- W^x_{i+1}, W^x_{j-1} are simple terms. Then the substring $W^x_{[i+1,j-1]}$ is of the form **(M3)**, it is shorter than $W^x_{[i,j]}$ hence $M(i+1, j-1) = 1$. By the induction hypothesis that string is found by the algorithm correctly so $M(i,j) = 1$ (case **(A2)**).

- $W^x_{i+1} = *, W^x_{j-1} \in T$. So $W^x_{[i,j]}$ is of the form $p^{(m)} * ...\rangle...\langle...q^{(m+1)}$. Let j' be the position of the first left angle bracket in $W^x_{[i,j]}$. We know there exists $j' < k < j - 1$ such that $W^x_k W^x_{j-1}$ reduces to 1. Hence $W^x_{[j',j-1]}$ is of the form **(M2a)**, it is shorter than $W^x_{[i,j]}$ hence $M(j', j - 1) = 1$. By the induction hypothesis that string is found by the algorithm correctly so $M(i,j) = 1$ (case **(A4a)**).

- $W^x_{i+1} \in T, W^x_{j-1} = *$. So $W^x_{[i,j]}$ is of the form $p^{(m)}...\rangle...\langle... * q^{(m+1)}$. Let i' be the position of the last right angle in $W^x_{[i,j]}$. We know there exists $i + 1 < k < i'$ such that $W^x_{i+1} W^x_k$ reduces to 1. Hence $W^x_{[i+1,i']}$ is of the form **(M2b)**, it is shorter than $W^x_{[i,j]}$ hence $M(i + 1, i') = 1$. By the induction hypothesis that string is found by the algorithm correctly so $M(i,j) = 1$ (case **(A4b)**).

- $W^x_{i+1} = *, W^x_{j-1} = *$. Then the string $W^x_{[i,j]}$ can be of the form $p^{(m)} * U\rangle\langle V * q^{(m+1)}$ and then $M(i,j) = 1$ by the initial case of the description of the algorithm. If $W^x_{[i,j]}$ contains more brackets, then there is a string $W^x_{[k,k']}$, where k is the index of the first left angle bracket in the string $W^x_{[i,j]}$ and k' is the index of the last right angle bracket in the string $W^x_{[i,j]}$. $W^x_{[k,k']}$ is of the form **(M4)**, it is shorter than $W^x_{[i,j]}$ therefore by the induction hypothesis $M(k, k') = 1$. So $M(i,j) = 1$ by **(A4c)**.

- 4. $W^x_{[i,j]}$ is of the form **(M4)**. Then let k be the index of the first simple term that participates in the reduction (that is the first simple term in the substring $\mathbb{D}_1$, so it is obviously in between the first pair of matched angle brackets) and k' be the index of the last simple term that participates in the reduction (that is the last simple term in the substring $\mathbb{D}_g$, so it is obviously in between the last pair of matched angle brackets). The substring $W^x_{[k,k']}$ is of the form **(M3)** is shorter than $W^x_{[i,j]}$ so by induction hypothesis $M(k, k') = 1$. Therefore $M(i,j) = 1$ by **(A5)**.

The algorithm is polynomial. One can observe that all conditions for $W^x_{[i,j]}$ can be checked in $O(j - i)$ steps. The number of the substrings is equal to $O(|W^x|^2)$. So one can compute the function $M(1, |W^x|)$ in time $O(|W^x|^3)$.

3.3 Obtaining the Reduction

The algorithm can be easily modified to become a parsing algorithm. Each obtained reduction is described by the set of links that take part in it. If we want to obtain only one reduction the complexity of the algorithm does not increase. The set of links $L(i, j)$ represents a reduction of some term to 1. Links are denoted by pairs of integers (k, l) such that $i \leqslant k < l \leqslant j$. We find the set of links by backtracking the indices of the function $M(i, j) = 1$ obviously starting with $M(1, |W^x|)$. We also define an auxiliary function $Prev(i, j)$ to help us follow the backtracking (as the value of the function $M(i, j)$ does not tell how it was obtained). The value of the function $Prev(i, j)$ is a sequence of three pairs $((l_1, l_2), (m1_1, m1_2), (m2_1, m2_2))$ where l_1, l_2 are indices of the link, $m1_1, m1_2, m2_1, m2_2$ are indices of function $\mathbf{M}$ on which the computation $M(i, j) = 1$ is based. If any of the values is not used it is set to 0. Every time when the algorithm computes the value of the function $M(i, j) = 1$ we set the value of the function $Prev(i, j)$ in the following way:

- for any initial case, $Prev(i, j) = ((i, j), (0, 0), (0, 0))$
- if it is a computation by one of the cases: (**A2**), (**A4a**), (**A4b**), (**A4c**), then $Prev(i, j) = ((i, j), (k, l), (0, 0))$, where (k, l) is the pair of indices for which the value the function M was 1 in the current computation (that is e.g. in (**A2**) a pair $(k, l) = (i + 1, j - 1)$)
- if it is a computation by one of the cases: (**A3a**), (**A3b**), (**A5**), then $Prev(i, j) = ((0, 0), (k, l), (0, 0))$ where (k, l) is the pair of indices for which the value the function M was 1 in the current computation
- if it is a computation by one of the cases: (**A1a**), (**A1b**), then $Prev(i, j) = ((0, 0), (i, k), (k + 1, j))$

When the computation of the functions M and $Prev$ is finished we easily compute the set $L(1, |W^x|)$. The definition of the function $L(i, j)$ is as follows:

- if $Prev(i, j) = ((i, j), (0, 0), (0, 0))$, where $0 < i < j$, then $L(i, j) = \{(i, j)\}$
- if $Prev(i, j) = ((i, j), (k, l), (0, 0))$, where $0 < i \leqslant k < l \leqslant j$, then $L(i, j) = L(k, l) \cup \{(i, j)\}$
- if $Prev(i, j) = ((0, 0), (k, l), (0, 0))$, where $0 < i \leqslant k < l \leqslant j$, then $L(i, j) = L(k, l)$
- if $Prev(i, j) = ((0, 0), (i, k), (k + 1, j))$, where $0 < i < k < j$, then $L(i, j) = L(i, k) \cup L(k + 1, j)$.

4   Examples

We run our algorithm on some linguistic examples from [6], [8].

Example 1. Let us take a sample dictionary:

I: π_1
will: $\pi^r s_1 j^l,\ q_1 j^l \pi^l$
meet: io^l
him: o

with

$$\pi_1 \leqslant \pi,\quad i \leqslant j,\quad s_1 \leqslant s$$

Let us consider the sentence:

I will meet him.

We construct a string W_x:

$$\langle *\pi_1^{(0)} *\rangle \langle *\pi^{(1)} s_1^{(0)} j^{(-1)} *q_1^{(0)} j^{(-1)} \pi^{(-1)} *\rangle \ \langle * \ i^{(0)} \ o^{(-1)} \ *\rangle \ \langle \ * \ o^{(0)} \ * \ \rangle \ \langle \ * \ s^{(1)}$$
$$\begin{array}{cccccccccccccccccccccccccccccc} 1&2&3&&4&5&6&7&8&&9&&10&11&12&13&&14&&15&16&17&18&19&&20&&21&22&23&24&25&26&27&28&29&30 \end{array}$$

We can compute the function $M(i, j)$. There are no substrings of the length 2 such that $M(i, j) = 1$. We start with the following $M(i, j)$:

1. $M(3, 8) = 1$
2. $M(10, 19) = 1$
3. $M(20, 25) = 1$

Then we compute:

4. $M(10, 25) = 1$ by the lines 2,3 and the case (**A1a**)
5. $M(10, 27) = 1$ by the line 4 and the case (**A3b**)
6. $M(9, 30) = 1$ by the line 5 and the case (**A4b**)
7. $M(3, 30) = 1$ by the lines 1,6 and the case (**A1a**)
8. $M(1, 30) = 1$ by the line 7 and the case (**A3a**)

Therefore there exists a type assignment for a given string of words which reduces to s.

The output of the parsing algorithm is:
$\{(3,8),\ (9,30),\ (10,19)\ (20,25)\}$

Example 2. The dictionary:

did: $q_2 i^l \pi^l,\ \pi^r s_2 i^l$
he: π_3
give: $io^l,\ i\overline{n}^l o'^l$
books: n
to: $i^r io^l,\ \overline{j} i^l$
her: o

with

$$\pi_j \leqslant \pi,\ s_i \leqslant s,\ q_k \leqslant q \leqslant s,\ \overline{j} \leqslant \pi_3,\ o' \leqslant o,\ n \leqslant \overline{n} \leqslant o$$

We consider the sentence:

Did he give books to her?

We construct a string W_x:

$< *\pi^r s_2 i^l * q_2 i^l \pi^l * >< *\pi_3* >< * io^l * i\overline{n}^l o''^l * >< *n* >< *i^r io^l * \overline{j} i^l * >< *o* >< *s^r$
$1 \qquad\quad 7\ 8\ 9 \qquad\quad 14\ 16 \quad 19\ 20 \qquad\qquad 29 \qquad 34\ 36 \qquad\qquad 44 \qquad 49$

We can compute the function $M(i,j)$. There are no substrings of the length 2 such that $M(i,j) = 1$. The algorithm computes $M(i,j) = 1$:

1. $M(9,14) = 1$ by initial case
2. $M(9,16) = 1$ by line 1 and (**A3b**)
3. $M(8,19) = 1$ by line 2 and (**A4b**)-useless
4. $M(8,22) = 1$ by line 2 and (**A4b**) - useless
5. $M(20,29) = 1$ by initial case
6. $M(8,29) = 1$ by lines 3,5 and (**A1a**)-useless
7. $M(20,31) = 1$ by line 5 and (**A3b**)
8. $M(8,31) = 1$ by line 6 and (**A3b**)-useless
9. $M(19,34) = 1$ by line 7 and (**A4b**)
10. $M(17,34) = 1$ by line 9 and (**A4a**)
11. $M(9,34) = 1$ by lines 2,10 and (**A1b**)
12. $M(8,35) = 1$ by line 11 and (**A2**)
13. $M(36,44) = 1$ by initial case
14. $M(8,44) = 1$ by lines 12,13 and (**A1a**)
15. $M(36,46) = 1$ by line 13 and (**A3b**)-useless
16. $M(8,46) = 1$ by line 14 and (**A3b**)
17. $M(7,49) = 1$ by line 16 and (**A4b**)
18. $M(1,49) = 1$ by line 17 and (**A3a**)

Therefore there exists a type assignment for a given string which reduces to s. The output of the parsing algorithm is:
$\{(7,49), (8,35), (9,14), (19,34), (20,29), (36,44)\}$

Example 3. An example from [8].
The dictionary:

Kim: np
mailed: $np^r\ s\ pp^l\ np^l, np^r\ s\ np^l$
the: $np\ n^l$
letter: $n\ pp^l,\ n$
to: $pp\ np^l$
Sandy: np

We consider the sentence:
Kim mailed the letter to Sandy.
We construct a string W_x:

$< *np* >< *np^r\ s\ pp^l\ np^l * np^r\ s\ np^l * >< *np\ n^l * >< *n\ pp^l * n* >$
$1 \quad 3 \qquad\quad 8\ 9\ 10\ 11 \qquad\qquad 17 \quad 20\ 21 \qquad\qquad 29$

$$< *pp\ np^l* >< *np* >< *s^r$$
$$32\ \ 34\ 35 \qquad\quad 40 \qquad\quad 45$$

1. $M(3,8) = 1$ by initial case -(r1)
2. $M(1,8) = 1$ by line 1 and (**A3a**) - useless
3. $M(3,13) = 1$ by initial case -(r2)
4. $M(1,13) = 1$ by line 3 and (**A3a**) - useless
5. $M(15,20) = 1$ by initial case - (r2)
6. $M(11,20) = 1$ by initial case - (r1)
7. $M(21,26) = 1$ by initial case - (r2)
8. $M(15,26) = 1$ by lines 5,7 and (**A1a**) - (r2')
9. $M(11,26) = 1$ by line 6,7 and (**A1a**) - useless
10. $M(21,29) = 1$ by initial case - (r1)
11. $M(15,29) = 1$ by lines 5,10 and (**A1a**)
12. $M(11,29) = 1$ by lines 6,10 and (**A1a**)- (r1)
13. $M(21,31) = 1$ by line 10 and (**A3b**) - useless
14. $M(15,31) = 1$ by line 11 and (**A3b**) - useless
15. $M(11,31) = 1$ by line 12 and (**A3b**) - (r1)
16. $M(27,34) = 1$ by initial case - (r2)
17. $M(21,34) = 1$ by lines 7,16 and (**A1a**) - (r2")
18. $M(15,34) = 1$ by lines 5,17 and (**A1a**) or by line 8,16 and (**A1a**) - both leading to (r2)
19. $M(11,34) = 1$ by lines 6,17 and (**A1a**) - useless
20. $M(10,34) = 1$ by line 15 and (**A4b**) - (r1)
21. $M(35,40) = 1$ by initial case - (r1) and (r2)
22. $M(27,40) = 1$ by lines 16,21 and (**A1a**) - (r2)
23. $M(21,40) = 1$ by lines 7,22 and (**A1a**) - (r2) or by lines 17,20 and (**A1a**)
24. $M(15,40) = 1$ by lines 5,23 and (**A1a**) - (r2) or by lines 8,22 - (r2') and (**A1a**) or by lines 18,21 and (**A1a**) - (r2")
25. $M(11,40) = 1$ by lines 6,23 and (**A1a**) - useless
26. $M(10,40) = 1$ by lines 20,21 and (**A1a**) - (r1)
27. $M(35,42) = 1$ by line 21 and (**A3b**) - useless
28. $M(27,42) = 1$ by line 22 and (**A3b**) - useless
29. $M(21,42) = 1$ by line 23 and (**A3b**) - useless
30. $M(15,42) = 1$ by line 24 and (**A3b**) - (r2)
31. $M(11,42) = 1$ by line 25 and (**A3b**) - useless
32. $M(10,42) = 1$ by line 26 and (**A3b**) - (r1)
33. $M(14,45) = 1$ by line 30 and (**A4b**) - (r2)
34. $M(9,45) = 1$ by line 32 and (**A4b**) - (r1)
35. $M(3,45) = 1$ by lines 1,34 and (**A1a**) - (r1) or lines 3,33 and (**A1a**) - (r2)
36. $M(1,45) = 1$ by line 35 and (**A3a**) - (r1) and (r2)

There are 2 possible reductions:

$\{(3,8), (9,45), (10,34), (11,20), (21,29), (35,40)\}$ and
$\{(3,13), (14,45), (15,20), (21,26), (27,34), (35,40)\}$

5   Algorithms of Oehrle and Savateev

For a better readability of this paper we briefly discuss related algorithms of
Oehrle [8] and Savateev [10].

The run of Oehrle's algorithm on the last example is presented on pp. 67-71
of [8], so it takes much more space than our presentation. Oehrle's algorithm
is essentially different. It represents the entry $a_1...a_n$ as a directed graph. Each
type A assigned to a_i by G is represented as a linear graph of simple terms; one
draws arcs from the source 1 to the first simple terms of these types and from the
last simple terms of these types to the target 1. Then one identifies the target
of the graph of a_i with the source of the graph of a_{i+1}. The algorithm adds new
arcs to the resulting graph. For any path $t_1 \rightarrow 1 \rightarrow t_2$, one adds $t_1 \rightarrow t_2$, and
for any path $t_1 \rightarrow p^{(}n) \rightarrow p^{(}n+1) \rightarrow t_2$, one adds $t_1 \rightarrow t_2$ ([8] does not regard
poset rules). This procedure is recursively executed, running the graph from the
left to the right. At the end, it returns the so-called winners, i.e. simple terms t
such that the final graph contains the path $1 \rightarrow t \rightarrow 1$ from the source of a_1 to
the target of a_n.

Oehrle [8] shows that the time is $O(n^3)$, but the size of G is treated as a
constant. It can be shown that the time is cubic also in our sense (depending on
$|W^x|$). However, the work of the algorithm seems more complicated than ours;
at each step, it has to remember all predecessors and successors of every vertice
and update some of them. Also, our algorithm can easily be modified to admit
an arbitrary type $\mathbb{B}$ in the place of s: it suffices to replace $s^{(1)}$ by $\mathbb{B}^r$, where
$(p_1^{(n_1)}...p_k^{(n_k)})^r = p_k^{(n_k+1)}...p_1^{(n_1+1)}$. This generalization is not directly applicable
to Oehrle's algorithm.

The algorithm of Savateev [10] is a recognition algorithm for $\mathbf{L}^\backslash$-grammars.
Recall that $\mathbf{L}^\backslash$ is the \-fragment of the Lambek calculus. Types are formed out
of atomic types $p, q, r, ...$ by the rule: if A, B are types, then $(A\backslash B)$ is a type. An
$\mathbf{L}^\backslash$-grammar assigns a finite number of types to each $a \in \Sigma$. $L(G)$ is defined as
for pregroup grammars except that $\mathbf{CBL}$ is replaced by $\mathbf{L}^\backslash$ (a comma separates
types, whereas for pregroup grammars it is interpreted as the concatenation of
strings).

Atoms are expressions p^n, where p is an atomic type and n is a positive
integer. For each type A, $\gamma(A)$ is a finite string of atoms, defined as follows:
$\gamma(p) = p^1$, $\gamma(A\backslash B) = (\gamma(A))^\top \gamma(B)$, where $(p^{n_1}...p^{n_k})^\top = p^{n_k+1}...p^{n_1+1}$. This is
precisely the translation of types in the formalism of free pregroups except that
Savateev writes p^{n+1} for our $p^{(n)}$.

A string of atoms has a good pairing, if there exists a system of non-crossing
links (as in pregroup reductions) such that:

(1) each atom is a left or right end of exactly one link,
(2) if p^n is the left end of a link, then its right end is p^{n+1},

(3) if p^n is the left end of a link and n is even, then the interval between p^n and the right end p^{n+1} contains an atom p^m with $m < n$.

The key lemma states that $A_1, ..., A_n \to B$ is provable in $\mathbf{L}^{\backslash}$ if and only if the string $\gamma(A_1)...\gamma(A_n)(\gamma(B))^{\top}$ has a good pairing. Clearly (1) and (2) yield the reducibility of the string to 1 in $\mathbf{CBL}$ (after replacing p^{n+1} by $p^{(n)}$), and (3) is an additional constraint (necessary, since $\mathbf{L}^{\backslash}$ is weaker than $\mathbf{CBL}$).

The recognition algorithm for $\mathbf{L}^{\backslash}$-grammars represents the entry $x = a_1...a_n$ like in our algorithm; each type A assigned to a_i is replaced by $\gamma(A)$. The algorithm checks whether there exists a string $A_1, ..., A_n$ of types assigned to $a_1, ..., a_n$, respectively, such that the string $\gamma(A_1)...\gamma(A_n)(\gamma(B))^{\top}$ has a good pairing. (B is the designated type of G.)

We will not write details of the algorithm and the proof of correctness, given in [10]. As mentioned in Section 1, although the algorithm is generally similar to ours (we follow it!), several details are essentially different. For instance, Savateev's algorithm and the proof of its correctness rely upon the fact that the right-most atom in $\gamma(A)$ has the form p^1, and other peculiarities of the translation. Thus, Savateev's algorithm does not correctly check whether an arbitrary string of atoms has a good pairing; it is correct only for strings which are related to $\mathbf{L}^{\backslash}$-sequents in the way described above. Our algorithm handles arbitrary strings of terms, admits both positive and negative integers n (i.e. both right and left adjoints), and needs no constraint like (3).

6   Conclusion

We have presented a new cubic, dynamic parsing algorithm for ambiguous pregroup grammars. The algorithm modifies essentially the recognition algorithm of Savateev [10], which was defined for categorial grammars based on $\mathbf{L}^{\backslash}$. The algorithm works on a string, of a specified construction, containing all possible types assigned to each word of the string to parse and it checks if a given string of words is a sentence. In case of a positive answer it returns one of possible reductions to the sentence type. We have proved the correctness of the algorithm. We have given a few examples on the run of the algorithm. Finally, we have compared our algorithm with the algorithm of Oehrle [8] for pregroup grammars and the algorithm of Savateev [10].

References

1. Béchet, D.: Parsing Pregroup Grammars and Lambek Calculus using Partial Composition. Studia Logica 87(2-3), 199–224 (2007)
2. Buszkowski, W.: Lambek Grammars Based on Pregroups. In: de Groote, P., Morrill, G., Retorè, C. (eds.) LACL 2001. LNCS (LNAI), vol. 2099, p. 95. Springer, Heidelberg (2001)
3. Buszkowski, W.: Type Logics and Pregroups. Studia Logica 87(2-3), 145–169 (2007)
4. Buszkowski, W., Moroz, K.: Pregroup grammars and context-free grammars. In: Casadio, C., Lambek, J. (eds.) Computational Algebraic Approaches to Natural Language, Polimetrica, pp. 1–21 (2008)

5. Lambek, J.: Type Grammars Revisited. In: Lecomte, A., Lamarche, F., Perrier, G. (eds.) LACL 1997. LNCS (LNAI), vol. 1582, pp. 1–27. Springer, Heidelberg (1999)

6. Lambek, J.: From Word to Sentence: a computational algebraic approach to grammar, Polimetrica (2008)

7. Mootrgat, M., Oehrle, R.: Pregroups and type-logical grammar: searching for convergence. In: Casadio, C., Scott, P.J., Seely, R.A.G. (eds.) Language and Grammar, pp. 141–160. CSLI Publications, Stanford (2005)

8. Oehrle, R.: A parsing algorithm for pregroup grammars. In: Proceedings of Categorial Grammars, Montpellier France, pp. 59–75 (2004)

9. Preller, A.: Linear Processing with Pregroups. Studia Logica 87(2-3), 171–197 (2007)

10. Savateev, Y.: Unidirectional Lambek Grammars in Polynomial Time, Theory of Computing Systems (to appear)

Term Graphs and the NP-Completeness of the Product-Free Lambek Calculus

Timothy A.D. Fowler

University of Toronto
tfowler@cs.toronto.edu

Abstract. We provide a graphical representation of proofs in the product-free Lambek calculus, called term graphs, that is related to several other proof net presentations. The advantage of term graphs is that they are very simple compared to the others. We use this advantage to provide an NP-completeness proof of the product-free Lambek Calculus that uses the reduction of [1]. Our proof is more intuitive due to the fact that term graphs allow arguments that are graphical in nature rather than using the algebraic arguments of [1].

1 Introduction

The Lambek calculus [2], is a variant of categorial grammar that is of interest to computational linguists because of its ability to capture a wide range of the semantics of natural language. In this paper we will only be concerned with the product-free fragment because of its simplicity and the paucity of linguistic uses of the product connective. The product-free Lambek calculus has a number of interesting computational properties. First, it is weakly equivalent to context-free grammars [3] but not strongly equivalent [4]. Second, it has recently been proven to have an NP-complete sequent derivability problem [1]. Finally, recent research has shown that as long as the order of categories is bounded, polynomial time parsing is possible [5,6].

With these latter two results, we have precisely determined where the intractability of the Lambek calculus lies, but the NP-completeness proof of [1] and the polynomial time algorithm of [6] seem to use entirely different methods for proving correctness. The former uses a primarily algebraic approach whereas the latter uses a graphical approach based on the graphical LC-Graphs of [7].

The purpose of this paper will be to introduce a new representation of proof nets, called *term graphs*[1], that are similar to the LC-Graphs of [7] but simpler in that they require two less correctness conditions and they avoid the introduction of terms from the lambda calculus. Term graphs are important because they bridge the gap between the methods of [1] and [6]. That is, despite being superficially very different from the

---

[1] A reviewer claimed that the correct reference for these graphs is a technical report by Lamarche in 1994. There is no such technical report, but there does seem to have been a manuscript that was circulated to several of Lamarche's colleagues. To our knowledge, the first publication of these criteria was by [8], which occurred after both the presentations of [9] and [10] at conferences. The first correctness proof was published by [7]. In response to the present author's request for the 1994 manuscript, it was released as an INRIA technical report [11].

P. de Groote, M. Egg, and L. Kallmeyer (Eds.): FG 2009, LNAI 5591, pp. 150–166, 2011.
© Springer-Verlag Berlin Heidelberg 2011

structures of [1], they are fundamentally quite similar, as we will see. In addition, the *abstract term graphs* of [6] are an abstraction over graphical structures that are essentially identical to our term graphs. Once we have introduced term graphs, we will use them to provide a proof of the NP-completeness of the sequent derivability problem for the product-free Lambek calculus that is more intuitive than that of [1] giving us insight into his reduction. This NP-completeness proof also allows us to consider these two results in the same language which will help future research in the area.

This paper will proceed as follows: In section 2, we will introduce the Lambek calculus and term graphs and prove the correctness of term graphs. Then, in section 3 we will introduce the polynomial reduction of [1] and provide an NP-completeness proof that is graphical in nature.

2   The Lambek Calculus and Term Graphs

In this section, we introduce the Lambek calculus in its sequent presentation and then quickly ignore the sequent presentation in favour of term graphs. Term graphs, like other proof net methods, allow easier analysis of computational problems pertaining to these sorts of logics. We prove the correctness of term graphs via the LC-graphs of [7].

2.1   The Lambek Calculus

The sequent derivability problem for the Lambek calculus takes as input a sequent made up of *categories*. The categories for the product-free fragment are built up from a set of *atoms* and the two binary connectives $/$ and $\backslash$. A sequent is a sequence of categories known as the antecedent together with the $\vdash$ symbol and one additional category called the succedent. The sequent derivability problem asks whether an input sequent is logically derivable from the axioms and rules shown in figure 1.

We will be considering two closely related variants of the Lambek calculus: The original Lambek calculus (L) and the Lambek calculus allowing empty premises (L^*). A sequent is *derivable in L^** if and only if it has a proof according to the sequent calculus in figure 1. In addition, we say that the sequent is derivable in L if and only if it is derivable in L^* such that Γ is non-empty when applying the rules $\backslash R$ and $/R$.

In figure 1, lowercase Greek letters represent categories and uppercase Greek letters represent sequences of categories.

$$\frac{}{\alpha \vdash \alpha}$$

$$\frac{\Gamma \vdash \alpha \qquad \Delta\beta\Theta \vdash \gamma}{\Delta\Gamma\alpha\backslash\beta\Theta \vdash \gamma} \backslash L \qquad \frac{\alpha\Gamma \vdash \beta}{\Gamma \vdash \alpha\backslash\beta} \backslash R$$

$$\frac{\Gamma \vdash \alpha \qquad \Delta\beta\Theta \vdash \gamma}{\Delta\beta/\alpha\Gamma\Theta \vdash \gamma} / L \qquad \frac{\Gamma\alpha \vdash \beta}{\Gamma \vdash \beta/\alpha} / R$$

Fig. 1. Axioms and rules of the Lambek calculus (from [2])

2.2   Term Graphs

In this section, we will introduce *term graphs*[2] which are a simplification of the LC-Graphs of [7], which in turn are based on the proof nets of [12]. The advantage of term graphs over LC-Graphs is that we have only two correctness conditions instead of four and the fact that we avoid the introduction of lambda terms. Furthermore, both of the term graph correctness conditions are conditions on the existence of certain paths whereas the LC-Graph correctness conditions are conditions on the existence *and* absence of certain paths.

Definition 1. *A* term graph *for a sequent is a directed graph whose vertices are category occurrences and whose edges are introduced in four groups. Like other proof net presentations, we will proceed with a deterministic step first and a non-deterministic step second.*

First, we assign polarities to category occurrences by assigning negative polarity to occurrences in the antecedent and positive polarity to the succedent. Then, the first two groups of edges are introduced by decomposing the category occurrences via the following vertex rewrite rules:

$$(\alpha/\beta)^- \Rightarrow \alpha^- \to \beta^+ \tag{1}$$

$$(\beta\backslash\alpha)^- \Rightarrow \beta^+ \leftarrow \alpha^- \tag{2}$$

$$(\alpha/\beta)^+ \Rightarrow \beta^- \dashleftarrow \alpha^+ \tag{3}$$

$$(\beta\backslash\alpha)^+ \Rightarrow \alpha^+ \dashrightarrow \beta^- \tag{4}$$

Each vertex rewrite rule specifies how to rewrite a single vertex on the left side to two vertices on the right side. The neighbourhood of the vertex on the left side of each rule is assigned to α on the right side. Dashed edges are referred to as Lambek edges *and non-dashed edges are referred to as* regular edges. *These two groups of edges will also be referred to as* rewrite *edges.*

After decomposition via the rewrite rules, we have an ordered set of polarized vertices, with edges between some of them. We say that a vertex belongs *to a category occurrence in the sequent if there is a chain of rewrites going back from the one that introduced this vertex to the one that rewrote the category occurrence.*

Lemma 1. *After decomposition via the rewrite rules has terminated, there is a unique vertex with in-degree 0 belonging to each category occurrence in the sequent.*

Proof. *By induction over the rewrite rules.*

A third group of edges is introduced such that there is one Lambek edge from the unique vertex with in-degree 0 in the succedent to each unique vertex with in-degree 0 in each of the antecedent category occurrences. These edges are referred to as rooted Lambek edges. *This completes the deterministic portion of term graph formation.*

A matching *is a planar matching of these vertices in the half plane where atom occurrences are matched to atoms occurrences with the same atom but with opposite polarity.*

---

[2] We call them term graphs because they are a graphical representation of the semantic term.

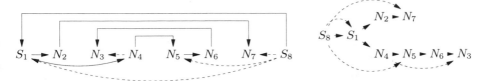

Fig. 2. Two depictions of an integral term graph for the sequent $(S/N)/(N/N), N/N, N \vdash S$. Polarities are not specified because they can be inferred due to lemma 3.

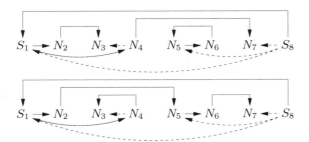

Fig. 3. An L^*-integral term graph that is not integral (top) and a term graph that is not L^*-integral (bottom) for the sequent $(S/N)/(N/N), N/N, N \vdash S$

The fourth group of edges are introduced as regular edges from the positive vertices to the negative vertices they are matched to. If α and β are matched in a matching M then we write $M(\alpha, \beta)$. See figures 2 and 3 for an example.

The two edge types of a term graph induce two distinct graphs, the *regular term graph* and the *Lambek term graph*. We will prefix the usual graph theory terms with *regular* and *Lambek* to distinguish paths and edges in these graphs.

Lemma 2. *In a term graph, there is a unique vertex with in-degree 0.*

Proof. By lemma 1, each category in the antecedent and succedent has a unique vertex with in-degree 0 before the introduction of the rooted Lambek edges. However, after the introduction of those edges, the only vertex with in-degree 0 is the one in the succedent.

Definition 2. *In a term graph, the unique vertex of in-degree 0 is denoted by τ.*

Lemma 3. *The vertices in a term graph have the following restrictions on degree:*

	Negative vertices	Positive vertices $\neq \tau$	τ
regular in-degree	1	1	0
regular out-degree	Arbitrary	1	1
Lambek in-degree	1	0	0
Lambek out-degree	0	Arbitrary	Arbitrary

Proof. By induction on the term graph formation process.

Because of this result, we can determine the polarity of a vertex by its incident edges. Therefore, we will often simplify our diagrams by omitting polarities.

Definition 3. *We define two conditions on term graphs:*

> **T**: *For all Lambek edges* $\langle s, t \rangle$ *there is a regular path from s to t.*
> **T(CT)**: *For each Lambek edge* $\langle s, t \rangle$, *there exists a negative vertex x such that there is a regular path from s to x and there is no rewrite Lambek edge* $\langle s', x \rangle$ *such that there is a regular path from s to* s'.[3]

A matching and its corresponding term graph are L^*-*integral iff they satisfy* **T**. *A matching and its corresponding term graph are* integral *iff they satisfy* **T** *and* **T(CT)**.

A partial matching is a matching which matches a subset of the polarized vertices and a partial term graph is the term graph of a partial matching. We will extend the notions of integrity to partial matchings by requiring that the integrity conditions are true of Lambek edges whose source and target are matched in the matching. Then, we can prove that the union of two integral partial matchings is an integral partial matching by considering those Lambek edges with a source matched by one matching and a target matched by the other do not violate the integrity conditions.

We will prove that integrity corresponds to sequent derivability in section 2.3. As discussed in the introduction, there are a number of connections between term graphs and the structures of [1]. For example:

- The set of negative vertices in a term graph corresponds to the set $\mathcal{N}_W$ of [1].
- A matched edge from s to t in a term graph corresponds to $\pi(t) = s$ for $t \in \mathcal{N}_W$ of [1]
- A rewrite edge from s to t in a term graph corresponds to $\varphi(t) = s$ of [1].
- A regular edge from s to t in a term graph corresponds to $\psi(t) = s$ of [1].
- The requirement that that matchings be planar and be between like atoms of opposite polarity correspond to the first three correctness conditions of [1]. **T** corresponds to the fourth correctness condition and **T(CT)** corresponds to the fifth.

2.3   Term Graph Correctness

We will prove the correctness of term graphs with respect to the Lambek calculus via the LC-Graphs of [7]. Since LC-Graphs and term graphs are constructed using similar algorithms, we will define LC-Graphs in terms of how they differ from term graphs rather than from scratch.

LC-Graphs are graphs whose vertex set V is a set of lambda calculus variables, which are introduced during the equivalent of the rewrite rule process. During this process, atom occurrences are associated with lambda terms. The leftmost variable in each lambda term is a unique identifier for the atom. This correspondance between lambda variables and atoms establishes a correspondence between LC-Graphs and term graphs. In addition to this superficial difference, LC-Graphs differ from term graphs structurally in the following three ways:

---

[3] T(CT) requires that we distinguish rewrite Lambek edges from rooted Lambek edges in the representation. To avoid clutter, we will not mark this difference in our figures.

1. The lambda variables in an LC-Graph are locally rearranged relative to the corresponding atom occurrences in a term graph as seen in the mapping in figures 4 and 5.
2. LC-Graphs do not distinguish between Lambek edges and regular edges.
3. LC-Graphs do not introduce any equivalent to the rooted Lambek edges.

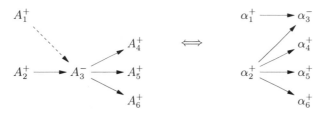

Fig. 4. The mapping between term graphs and LC-Graphs for neighbourhoods of negative vertices. The lambda variable α_i corresponds to the atom occurrence A_i for $1 \leq i \leq 6$.

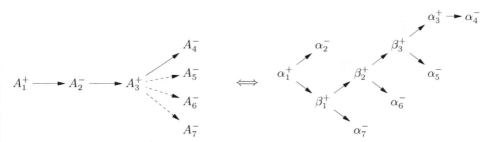

Fig. 5. The mapping between term graphs and LC-Graphs for neighbourhoods of positive vertices. The positive vertex A_3 in a term graph is represented by the vertices $\beta_1, \beta_2, \beta_3$ and α_3 in an LC-Graph. The lambda variable α_i corresponds to the atom occurrence A_i for $1 \leq i \leq 7$.

The first of these three differences is required for term graphs to avoid the introduction of lambda terms. The second difference allows us to express our correctness conditions more concisely and simplifies the presentation of our proofs. This is accomplished by no longer needing to identify the edges in an LC-Graph that are the equivalent of Lambek edges by their endpoint vertices. The last difference will allow us to eliminate one of the correctness conditions.

[7] defines the following terms, necessary to understand the correctness conditions:

Definition 4. *A* lambda-node *is a positive vertex in an LC-Graph with two regular out-neighbours, one of which is positive and one of which is negative. The positive one is its* plus-daughter *and the negative one is its* minus-daughter.

For example, B_1, B_2 and B_3 in figure 5 are lambda-nodes. The intuition is that they correspond to Lambek edges. We can now define the correctness conditions on LC-Graphs and state theorem 1 (proven in [7]):

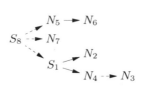

Fig. 6. The tree of rewrite and rooted Lambek edges for $(S/N)/(N/N), N/N, N \vdash S$

I(1): There is a unique node in G with in-degree 0, from which all other nodes are path-accessible.
I(2): G is acyclic.
I(3): For every lambda-node $v \in V$, there is a path from its plus-daughter u to its minus-daughter w
I(CT): For every lambda-node $v \in V$, there is a path in G, $v \rightsquigarrow x$ where x is a terminal node and there is no lambda-node $v' \in V$ such that $v \rightsquigarrow v' \rightarrow x$.

Theorem 1. *A sequent is derivable in L^* iff it has an LC-Graph satisfying conditions I(1-3). A sequent is derivable in L iff it has an LC-Graph satisfying conditions I(1-3) and I(CT).*

The equivalence between the **I** conditions and the **T** conditions begins with the following two lemmas.

Lemma 4. *In a term graph, the rewrite edges and the rooted Lambek edges form a tree with τ as its root.*

Proof. By induction on the rewrite rules together with the way that the rooted Lambek edges are introduced.

Lemma 5. *In an L^*-integral term graph, if there is a path from a to b, then there is a regular path from a to b.*

Proof. The path from a to b may contain a Lambek edge $\langle s, t \rangle$, but by **T**, there is a regular path from s to t. Replacing the Lambek edge with the regular path and repeating gives us a regular path from a to b.

We next define a new condition on term graphs that is a counterpart to **I**(2).

T(C): The term graph is acyclic.

We now prove that any L^*-integral term graph is necessarily acyclic.

Proposition 1. $T \Rightarrow T(C)$

Proof. Suppose there is a cycle. Then, by lemma 5, there is a regular cycle. That cycle cannot contain τ since τ has no in-edges. However, by lemma 4, there is a path from τ to every vertex and by lemma 5 that path is regular. But, by lemma 3 all vertices have regular in-degree at most one. Therefore, no cycles can exist.

Proposition 2. $T \Rightarrow I(1)$

Proof. By lemma 4, there is a path from τ to every node in the term graph and by lemma 5 there is a regular path from τ to every node. Then, by the mapping of term graphs to LC-Graphs, these paths have equivalents in the LC-Graph.

Proposition 3. $T \Rightarrow I(2)$

Proof. By proposition 1, $\mathbf{T} \Rightarrow \mathbf{T}(C)$ and by inspection of the mapping from term graphs to LC-Graphs, no new cycles can be introduced.

Proposition 4. $T \Rightarrow I(3)$

Proof. By the mapping.

Proposition 5. $I(1), I(3) \Rightarrow T$

Proof. $\mathbf{I}(3)$ requires that all nodes be accessible from the root, which means that for the rooted Lambek edges $\langle \tau, t \rangle$, there is a regular path from τ to t. Then, because of the way that positive vertices in a term graph are mapped from their equivalents in LC-Graphs, enforcing $\mathbf{I}(3)$ requires that for rewrite Lambek edges $\langle s, t \rangle$, there be a regular path from s to t.

Proposition 6. $T(CT) \Leftrightarrow I(CT)$

Proof. The only point of interest is the fact that the Lambek edges whose source is τ could rule out some term graphs when are provable in L^*. However, such Lambek edges are specifically ruled out by $\mathbf{T}(CT)$.

Theorem 2. *A sequent is derivable in L iff it has a term graph satisfying T. A sequent is derivable in L^* iff it has a term graph satisfying conditions T and $T(CT)$.*

3   NP-Completeness Proof

Now that we have defined a simple graphical representation of proofs in the Lambek calculus, we can proceed with a graphical proof of NP-Completeness that in some ways mirrors the proof of [1]. We begin with the same reduction from SAT as [1].

Definition 5. *Let $c_1 \wedge \ldots \wedge c_m$ be a SAT instance with variables $x_1, \ldots, x_n$. We define the sequent Σ as in figure 7 (from [1]).*

The space of matchings is analyzed via two partial matchings M_t and N_t based on a truth assignment t (where truth assignments are sequences of booleans). First, we prove that the partial matching M_t is always integral and then we prove that the partial matching N_t is integral if and only if the SAT instance is satisfiable. Finally, in proposition 9, we prove that any L^*-integral matching must partition into two such partial matchings.

The atoms of Σ are $p_i^j, q_i^j, a_i^j, b_i^j, c_i^j$ and d_i^j for $1 \leq i \leq n$ and $1 \leq j \leq m$.

$$A_i^0 = a_i^0 \backslash p_i^0$$
$$A_i^j = (q_i^j/((b_i^j \backslash a_i^j) \backslash A_i^{j-1})) \backslash p_i^j$$
$$A_i = A_i^m$$
$$B_i^0 = a_i^0$$
$$B_i^j = q_{i-1}^j/(((b_i^j/B_i^{j-1}) \backslash a_i^j) \backslash p_{i-1}^j)$$
$$B_i = B_i^m \backslash p_{i-1}^m$$
$$G^0 = p_0^0 \backslash p_n^0$$
$$G^j = (q_n^j/((q_0^j \backslash p_0^j) \backslash G^{j-1})) \backslash p_n^j$$
$$G = G^m$$

$$C_i^0 = c_i^0 \backslash p_i^0$$
$$C_i^j = (q_i^j/((d_i^j \backslash c_i^j) \backslash C_i^{j-1})) \backslash p_i^j$$
$$C_i = C_i^m$$
$$D_i^0 = c_i^0$$
$$D_i^j = q_{i-1}^j/(((d_i^j/D_i^{j-1}) \backslash c_i^j) \backslash p_{i-1}^j)$$
$$D_i = D_i^m \backslash p_{i-1}^m$$
$$H_i^0 = p_{i-1}^0 \backslash p_i^0$$
$$H_i^j = ((q_{i-1}^j/(q_i^j/H_i^{j-1})) \backslash p_{i-1}^j) \backslash p_i^j$$
$$H_i = H_i^m$$

$$E_i^0(t) = p_{i-1}^0$$
$$E_i^j(t) = \begin{cases} q_i^j/(((q_{i-1}^j/E_i^{j-1}(t)) \backslash p_{i-1}^j) \backslash p_i^{j-1}) & \text{if } \neg_t x_i \text{ appears in } c_j \\ (q_{i-1}^j/(q_i^j/(E_i^{j-1} \backslash p_i^{j-1}))) \backslash p_{i-1}^j & \text{if } \neg_t x_i \text{ does not appear in } c_j \end{cases}$$
$$E_i(t) = E_i^m(t) \backslash p_i^m$$

$$\Pi_i = E_i(0)/(B_i \backslash A_i), H_i, (D_i \backslash C_i) \backslash E_i(1)$$
$$\Sigma = \Pi_1, \dots, \Pi_n \vdash G$$

Fig. 7. The sequent Σ for the SAT instance $c_1 \wedge \dots \wedge c_n$. Note that $\neg_0 x = \neg x$ and $\neg_1 x = x$

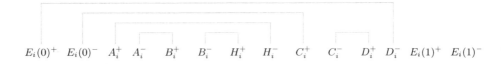

$$E_i(0)^+ \quad E_i(0)^- \quad A_i^+ \quad A_i^- \quad B_i^+ \quad B_i^- \quad H_i^+ \quad H_i^- \quad C_i^+ \quad C_i^- \quad D_i^+ \quad D_i^- \quad E_i(1)^+ \quad E_i(1)^-$$

Fig. 8. M_t for Π_i and $t = \langle t_1, \dots, t_{i-1}, 1, t_{i+1}, \dots, t_n \rangle$

Lemma 6. X has $4m + 2$ atoms for $X \in \{A_i, B_i, C_i, D_i, E_i(t), G, H_i\}$.

Proof. By induction.

Definition 6. For $X \in \{A_i, B_i, C_i, D_i, E_i(t), G, H_i\}$, X^+ is the leftmost $2m + 1$ atoms and X^- is the rightmost $2m + 1$ atoms. We refer to these as hills.

Proposition 7. Let $t = \langle t_1, \dots, t_n \rangle$ be a truth assignment and let M_t be the following partial matching (depicted in figure 8). For $1 \leq i \leq n$, $M_t(B_i^+, A_i^-)$ and $M_t(D_i^+, C_i^-)$. If $t_i = 0$ then $M_t(E_i(1)^+, B_i^-)$, $M_t(A_i^+, E_i(1)^-)$, $M_t(H_i^+, D_i^-)$ and $M_t(C_i^+, H_i^-)$. If $t_i = 1$ then $M_t(E_i(0)^+, D_i^-)$, $M_t(C_i^+, E_i(0)^-)$, $M_t(H_i^+, B_i^-)$ and $M_t(A_i^+, H_i^-)$.
  Then, M_t is integral.

Proof. Figure 9 shows the partial term graph consisting of the vertices from $B_i \backslash A_i$ (or equivalently from $D_i \backslash C_i$ which are identical under renaming of α and β). However, by introducing the edges from M_t for A_i, B_i, C_i and D_i, contracting paths regular path longer than 1 and removing Lambek edges $\langle s, t \rangle$ which have a regular path from s to t, we get the abstraction of a term graph shown in figure 10. This abstraction over a term graph is essentially identical to a term graph except that the Lambek in-degree of negative vertices is now unbounded due to path contraction.

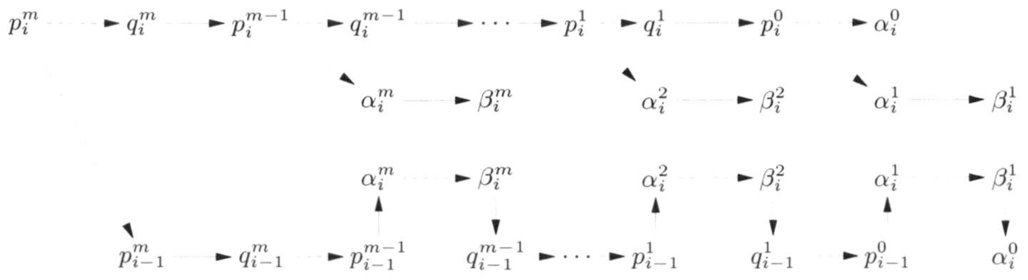

$$p_i^m \dashrightarrow q_i^m \longrightarrow p_i^{m-1} \longrightarrow q_i^{m-1} \longrightarrow \cdots \longrightarrow p_i^1 \longrightarrow q_i^1 \longrightarrow p_i^0 \dashrightarrow \alpha_i^0$$

$$\alpha_i^m \dashrightarrow \beta_i^m \qquad \alpha_i^2 \dashrightarrow \beta_i^2 \qquad \alpha_i^1 \dashrightarrow \beta_i^1$$

$$\alpha_i^m \dashrightarrow \beta_i^m \qquad \alpha_i^2 \dashrightarrow \beta_i^2 \qquad \alpha_i^1 \dashrightarrow \beta_i^1$$

$$p_{i-1}^m \longrightarrow q_{i-1}^m \longrightarrow p_{i-1}^{m-1} \qquad q_{i-1}^{m-1} \longrightarrow \cdots \longrightarrow p_{i-1}^1 \qquad q_{i-1}^1 \longrightarrow p_{i-1}^0 \qquad \alpha_i^0$$

Fig. 9. Term graph for $(B_i \backslash A_i)^+$ (for $\alpha = a$ and $\beta = b$) and $(D_i \backslash C_i)^+$ (for $\alpha = c$ and $\beta = d$)

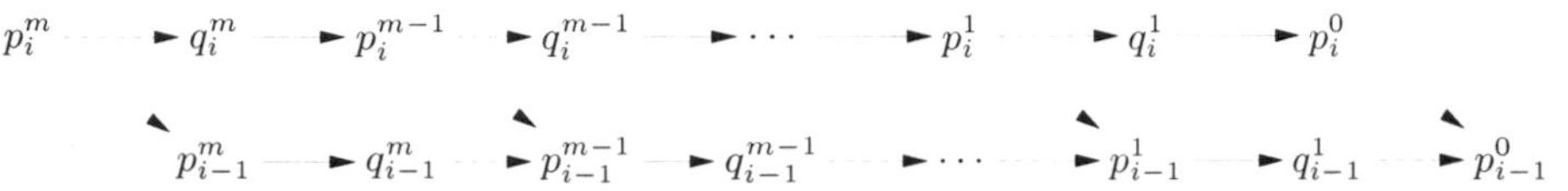

$$p_i^m \dashrightarrow q_i^m \longrightarrow p_i^{m-1} \longrightarrow q_i^{m-1} \longrightarrow \cdots \longrightarrow p_i^1 \longrightarrow q_i^1 \longrightarrow p_i^0$$

$$p_{i-1}^m \longrightarrow q_{i-1}^m \longrightarrow p_{i-1}^{m-1} \longrightarrow q_{i-1}^{m-1} \longrightarrow \cdots \longrightarrow p_{i-1}^1 \longrightarrow q_{i-1}^1 \longrightarrow p_{i-1}^0$$

Fig. 10. Term graph for $(B_i \backslash A_i)^+$ where B_i^- has been matched to A_i^+ (or $(D_i \backslash C_i)^+$ where D_i^- has been matched to C_i^+) and all paths have been contracted. The angled edges other than the first are abstracted edges similar to those in [6].

Then, regardless of whether $t_i = 0$ or $t_i = 1$, this partial abstract term graph is combined with the partial term graph for H_i (shown in figure 11) by inserting edges between identical atoms from the positive vertices to negative vertices. It can be seen that the combined term graph is L^*-integral by observing that each Lambek edge is in fact overlaid by a regular path.

In a parallel process, the abstract term graph for either $B_i \backslash A_i$ or $D_i \backslash C_i$ is combined with the term graph for $E_i(t_i)$, which is constructed out of components shown in figure 12. However, like the combination with H_i, the result is a term graph where all Lambek edges are forward edges despite the variation of $E_i(t_i)$ as exemplified in figure 13.

T(CT) is straightforward to check because for each Lambek edge $\langle s, t \rangle$ in these partial term graphs, the vertex s has a regular edge to a vertex x with a Lambek in-neighbour which does not have a regular path from s.

Therefore, M_t is integral.

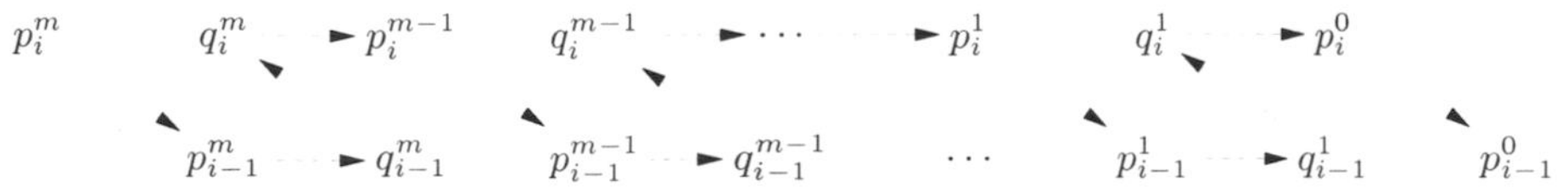

$$p_i^m \qquad q_i^m \searrow \; p_i^{m-1} \qquad q_i^{m-1} \searrow \; \cdots \longrightarrow p_i^1 \qquad q_i^1 \searrow \; p_i^0$$

$$p_{i-1}^m \longrightarrow q_{i-1}^m \qquad p_{i-1}^{m-1} \longrightarrow q_{i-1}^{m-1} \qquad \cdots \qquad p_{i-1}^1 \longrightarrow q_{i-1}^1 \qquad p_{i-1}^0$$

Fig. 11. Term graph for H_i

Proposition 8. *Let* $t = \langle t_1, \ldots, t_n \rangle$ *be a truth assignment and let* N_t *be the following partial matching (depicted in figure 14). For* $1 < i \leq n$, $N_t(E_i(t_i)^+, E_{i-1}(t_{i-1})^-)$, $N_t(E_1(t_1)^+, G^-)$ *and* $N_t(G^+, E_n(t_n)^-)$. *Then,* N_t *is integral iff* N_t *is* L^*-integral iff t *is a satisfying truth assignment for* $c_1 \wedge \ldots \wedge c_m$.

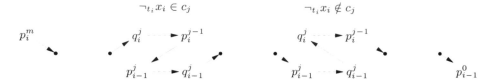

Fig. 12. Term graph components for $E_i(t_i)$. The round nodes are place holders used to indicate the source and target of some edges. They can be ignored once the term graph for $E_i(t_i)$ is complete.

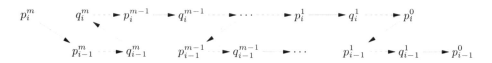

Fig. 13. Term graph for $E_2(0)$ for $(x_1 \vee x_2) \wedge (\neg x_2 \vee \neg x_1) \wedge (x_1) \wedge \ldots \wedge (x_2 \vee \neg x_1) \wedge (\neg x_2 \vee x_2)$

Proof. The relevant subgraph of the term graph for N_t is shown in figure 15 for the general case and in figure 16 for a specific example.

There are several types of Lambek pairs. Consider $\langle u, v \rangle$ such that $u, v \in E_i(t_i)$ for some i. Then, if $u = p_{i-1}^j$ and $v = q_{i-1}^j$ for $j \geq 1$, the path leaves u to the right but must eventually return to v since any rightward movement due to an F edge is mirrored by a leftward movement due its paired edge resulting in no **T** violation. If $u = q_i^j$ and $v = p_i^{j-1}$ for $j \geq 1$, either the link is completed immediately or it is completed via an angled edge whose target is v. The case where $u, v \in G$ where $u = p_n^j$ and $v = q_n^j$ is similar to the first case.

Next, consider $\langle u, v \rangle$ such that $u = p_i^{j-1}$ and $v = p_0^j$ for some $j \geq 1$. v is reached before u iff no edge T_i^j is present for $1 \leq i \leq n$ iff c_m does not contain $\neg_{t_i} x_i$ for any $1 \leq i \leq n$ iff $c_1 \wedge \ldots \wedge c_m$ is unsatisfiable.

Finally, consider the rooted Lambek edges $\langle u, v \rangle$ where $u = p_n^m$ and $v = p_i^m$ for some i. u is the leftmost atom in the bottom row and v appears somewhere to its right in the bottom row. Then, there must be a regular path from u to v since the angled edges target the lowest vertex without an in-edge.

Checking that **T**(CT) is never violated is straightforward via the same cases as above.

Definition 7. *Given a matching M, we denote the match of the occurrence of p_i^j in X^p to p_i^j in Y^q as $M(p_i^j, X^p, Y^q)$.*

Lemma 7. $M(p_i^m, E_i(0)^-, A_i^+)$ *cannot belong to an L^*-integral matching.*

$$E_1(t_1)^+ \quad E_1(t_1)^- \cdots E_2(t_2)^+ \quad E_2(t_2)^- \quad \cdots \quad E_n(t_n)^+ \quad E_n(t_n)^- \cdots \quad G^+ \qquad G^-$$

Fig. 14. N_t where $t = \langle t_1, t_2, \ldots, t_n \rangle$

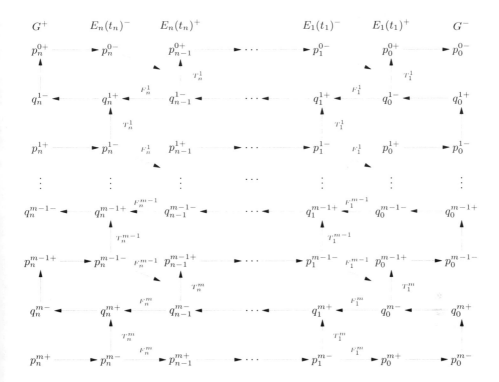

Fig. 15. The term graph N_t (with Lambek edges omitted). Each vertex in $\{p_i^{j-}, q_{i-1}^{j-}\}$ has two out-edges labelled T_i^j and F_i^j for $1 \leq j \leq m, 1 \leq i \leq n$. If $\neg_{t_i} x_i$ appears in c_j then we keep only the T_i^j edges and otherwise, we keep only the F_i^j edges. The angled edges, which appear to have no target, have a target which is dependant on whether $\neg_{t_k} x_k$ appears in c_l for $l < j$. Their target is always the lowest positive vertex which does not have another in-edge.

Proof. Because there is a regular edge from p_i^m in $E_i(0)^-$ to p_i^m in A_i^+ the matched edge would violate $T(C)$, as seen in figure 17.

Lemma 8. $M(p_{i-1}^m, D_i^-, E_i(1)^+)$ *cannot belong to an L^*-integral matching.*

Proof. Consider figure 18. If such an edge were to exist, then the only way for a regular path from p_i^m in C_i^+ to p_{i-1}^m in D_i^- to exist is for their to be a regular path between p_i^m in C_i^+ and some vertex in $E_i(1)$. However, the first vertex of that path among the vertices of $E_i(1)$ is the target of a Lambek edge $\langle s, t \rangle$ and a regular path from s to t would need to include p_i^m in $E_i(1)^-$. However, that would violate $T(C)$.

Proposition 9. *Any matching which does not extend M_t for some truth assignment t is not L^*-integral.*

Proof. Let M be an L^*-integral matching and let $1 \leq i \leq n$ be maximal such that neither of the following matchings are submatchings:

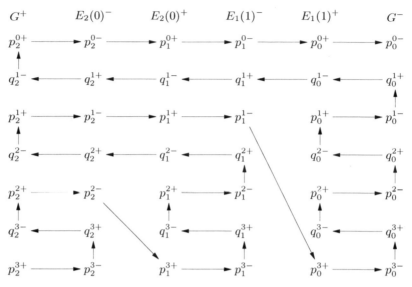

Fig. 16. The term graph N_t for the SAT instance $I = (\neg x_1 \vee x_2) \wedge (x_1 \vee x_2) \wedge (x_1 \vee \neg x_2)$ and the truth assignment $t = <0, 1>$. The fact that t is not a satisfying assignment for I corresponds to the fact that p_0^{1-} appears before p_2^{0+} in N_t.

$$E_i(0)^- \qquad\qquad\qquad\qquad A_i^+$$

$$p_i^0 \quad \cdots \quad q_i^{m-1} \quad p_i^{m-1} \quad q_i^m \quad p_i^m \longrightarrow p_i^m \longrightarrow q_i^m \longrightarrow p_i^{m-1} \longrightarrow q_i^{m-1} \longrightarrow \cdots \longrightarrow p_i^0$$

Fig. 17. A close-up of the boundary between $E_i(0)^-$ and A_i^+. The edges between the atoms of $E_i(0)^-$ have been omitted because they vary according to the SAT instance.

$$C_i^+ \qquad\qquad D_i^- \qquad\qquad E_i(1)^+ \qquad E_i(1)^-$$

$$p_i^m \searrow p_{i-1}^0 \longleftarrow \cdots \longleftarrow q_{i-1}^{m-1} \longleftarrow p_{i-1}^{m-1} \longleftarrow q_{i-1}^m \longleftarrow p_{i-1}^m \quad p_{i-1}^m \longleftarrow \cdots \longleftarrow \cdots \longleftarrow p_i^m$$

Fig. 18. A close-up of the boundary between D_i^- and $E_i(1)^+$. The edges in $E_i(1)$ vary depending on the SAT instance, but alternate regular and Lambek edges from p_i^m to p_{i-1}^m.

$$E_i(0)^+ \quad E_i(0)^- \quad A_i^+ \quad A_i^- \quad B_i^+ \quad B_i^- \quad H_i^+ \quad H_i^- \quad C_i^+ \quad C_i^- \quad D_i^+ \quad D_i^- \; E_i(0)^+ \quad E_i(0)^-$$

Fig. 19. The 14 hills of Π_i. The matches shown are obligatory because the atoms in those hills occur exactly twice in the term graph.

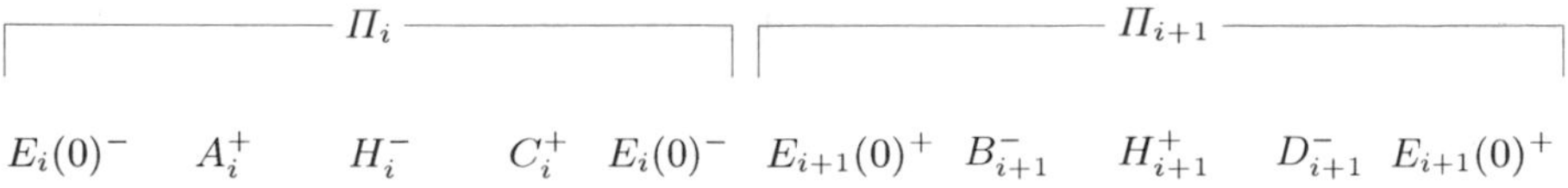

$$\overbrace{}^{\Pi_i} \qquad \overbrace{}^{\Pi_{i+1}}$$

$$E_i(0)^- \quad A_i^+ \quad H_i^- \quad C_i^+ \quad E_i(0)^- \quad E_{i+1}(0)^+ \quad B_{i+1}^- \quad H_{i+1}^+ \quad D_{i+1}^- \quad E_{i+1}(0)^+$$

Fig. 20. The 10 hills containing p_i^j for $i < n$

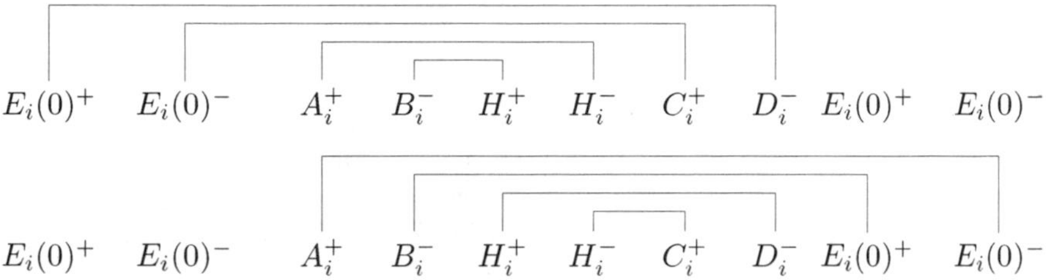

$$E_i(0)^+ \quad E_i(0)^- \quad A_i^+ \quad B_i^- \quad H_i^+ \quad H_i^- \quad C_i^+ \quad D_i^- \quad E_i(0)^+ \quad E_i(0)^-$$

$$E_i(0)^+ \quad E_i(0)^- \quad A_i^+ \quad B_i^- \quad H_i^+ \quad H_i^- \quad C_i^+ \quad D_i^- \quad E_i(0)^+ \quad E_i(0)^-$$

Let $0 \le j \le m$. Then, for $i = n$, p_i^j appears in 6 hills (from left to right):

$$E_i(0)^-, A_i^+, H_i^-, C_i^+, E_i(1)^-, G^+$$

For $i < n$, p_i^j appears in 10 hills:

$$E_n(0)^-, A_i^+, H_i^-, C_i^+, E_i(1)^-, E_{i+1}(0)^+, B_{i+1}^-, H_{i+1}^+, D_{i+1}^-, E_{i+1}(1)^+$$

However, because of our maximality assumption and the planarity requirement, we know that of the rightmost 5, only $E_{i+1}(t)^+$ does not match another of the rightmost 5. If $i = n$, let $E^+ = G^+$ and otherwise, let $E^+ = E_{i+1}(t)^+$. In either case, E^+ represents the rightmost unconstrained hill.

We now wish to consider the possible matchings of $E_i(0)^-, A_i^+, H_i^-, C_i^+, E_i(1)^-$ and E^+, but only for the occurrences of the atom p_i^j. For this section only, we will denote these matches by matching the hills they belong to, but we must remember that these matches are only for the p_i^j atoms and not the whole hill.

– Case 1: $\overbrace{E_i(0)^- \qquad A_i^+}$

Due to planarity, $M(p_i^j, E_i(0)^-, A_i^+)$ forces $M(p_i^m, E_i(0)^-, A_i^+)$ and by lemma 7 this matching is not L^*-integral.

– Case 2: $\overbrace{C_i^+ \qquad E_i(1)^-}$

Due to planarity, $M(p_i^j, C_i^+, E_i(1)^-)$ forces $M(p_{i-1}^m, D_i^-, E_i(1)^+)$ and by lemma 8 this matching is not L^*-integral.

– Case 3: $\overbrace{H_i^- \qquad E^+}$

Due to planarity, $M(p_i^j, H_i^-, E^+)$ forces $M(p_i^j, C_i^+, E_i(1)^-)$ which cannot be L^*-integral according to case 2.

This leaves us with only two possible matchings for p_i^j:

- Case 1: $E_i(0)^-$ A_i^+ H_i^- C_i^+ $E_i(1)^-$ E^+

  We will now shift from analyzing the matches for a general p_i^j and focus on one important atom. Consider the possible matches for the atom p_{i-1}^m in D_i^-, the rightmost atom in D_i^-. There are five, as can be seen in figure 20 (from left to right) which we will rule out:

  1. p_{i-1}^m in A_{i-1}^+

     Such a match is between the rightmost atom in D_i^- and the leftmost atom in A_{i-1}^+ as can be seen in figures 17 and 18. Then, we can see that no atom between these two has a regular out-edge to any atom not between these two. But, the regular in-neighbour of p_{i-1}^m in D_i^- is p_{i-1}^m in A_{i-1}^+ because of the match and the regular in-neighbour of p_{i-1}^m in A_{i-1}^+ is p_i^m in $E_{i-1}(0)^-$ because of the regular rewrite edge in figure 17. But, there is a Lambek edge whose target is p_{i-1}^m in D_i^- and whose source is p_i^m in C_i^+ as seen in figure 18. Then, there cannot possibly be a regular path from p_i^m in C_i^+ to p_{i-1}^m in D_i^-, resulting in a **T** violation.

  2. p_{i-1}^m in C_{i-1}^+

     Because of planarity, the atoms in D_{i-1}^- would need to match the atoms in $E_{i-1}(1)^+$. However, by lemma 8, M would not be L^*-integral.

  3. p_{i-1}^m in $E_i(0)^+$

     Contradicts our assumption that M does not have this submatching.

  4. p_{i-1}^m in H_i^+

     Such a match would violate planarity since p_i^j in $E_i(0)^-$ matches p_i^j in C_i^+.

  5. p_{i-1}^m in $E_i(1)^+$

     Then, M would not be L^*-integral by lemma 8.

- Case 2: $E_i(0)^-$ A_i^+ H_i^- C_i^+ $E_i(1)^-$ E^+

  As in the previous case, we will focus on one important atom. This time, that atom will be p_{i+1}^m in A_i^+, the leftmost atom in A_i^+. Again, there are five possible matches:

  1. p_{i+1}^m in $E_i(0)^-$

     Then, M would not be L^*-integral by lemma 7.

  2. p_{i+1}^m in H_i^-

     Such a match would violate planarity, since p_i^j in A_i^+ matches p_i^j in $E_i(1)^-$.

  3. p_{i+1}^m in $E_i(1)^-$

     Contradicts our assumption that M does not have this submatching.

  4. p_{i+1}^m in B_{i+1}^-

     Contradicts the maximality assumption (because B_{i+1}^- is part of Π_{i+1}).

  5. p_{i+1}^m in D_i^-

     Contradicts the maximality assumption (because D_{i+1}^- is part of Π_{i+1}).

Therefore, M must extend M_t for some truth assignment t.

Theorem 3. $c_1 \wedge \ldots \wedge c_m$ *is satisfiable iff* Σ *is derivable in* L *iff* Σ *is derivable in* L^*.

Proof. Propositions 7, 8 and 9 prove that any matching must partition into M_t and N_t for some truth assignment t. We need only consider the Lambek edges with a source in M_t and a target in N_t or vice versa but the only such edges are the rooted Lambek edges. It is tedious, but not difficult, to check that each such Lambek edge has an accompanying regular path and that $\mathbf{T}(CT)$ is not violated.

4   Conclusion

We have introduced a graphical representation of proof nets that is closely linked to both the structures of [1] and the abstract term graphs of [6]. Together these two results describe the boundary between tractability and intractability.

Our representation is very simple, requiring just two conditions (other than the matching conditions) to characterize correctness in the Lambek calculus. Furthermore, term graphs avoid the introduction of unneeded complexity such as the lambda terms of [7] and the algebraic terms of [13]. This has allowed us to provide a more intuitive proof of the NP-completeness result of [1], which allows us to more clearly see the boundary of tractability for the product-free Lambek calculus.

Acknowledgements

I must thank Gerald Penn for guiding this research as well as the useful comments of the three anonymous reviewers.

References

1. Savateev, Y.: Product-free lambek calculus is NP-complete. CUNY Technical Report (September 2008)
2. Lambek, J.: The mathematics of sentence structure. American Mathematical Monthly 65(3), 154–170 (1958)
3. Pentus, M.: Product-free lambek calculus and context-free grammars. The Journal of Symbolic Logic 62(2), 648–660 (1997)
4. Tiede, H.J.: Deductive Systems and Grammars: Proofs as Grammatical Structures. PhD thesis, Indiana University (1999)
5. Fowler, T.: Efficient parsing with the Product-Free lambek calculus. In: Proceedings of The 22nd International Conference on Computational Linguistics (2008)
6. Fowler, T.: A polynomial time algorithm for parsing with the bounded order lambek calculus. In: Ebert, C., Jäger, G., Michaelis, J. (eds.) MOL 10. LNCS, vol. 6149, pp. 36–43. Springer, Heidelberg (2010)
7. Penn, G.: A graph-theoretic approach to sequent derivability in the lambek calculus. Electronic Notes in Theoretical Computer Science 53, 274–295 (2004)
8. Perrier, G.: Intuitionistic multiplicative proof nets as models of directed acyclic graph descriptions. In: Nieuwenhuis, R., Voronkov, A. (eds.) LPAR 2001. LNCS (LNAI), vol. 2250, pp. 233–248. Springer, Heidelberg (2001)

9. Lamarche, F.: From proof nets to games. Electronic Notes in Theoretical Computer Science 3 (1996)
10. Penn, G.: A Graph-Theoretic approach to sequent derivability in the lambek calculus. In: Proceedings of Formal Grammar '01 and the 7th Meeting on Mathematics of Language. Helsinki, Finland (2001)
11. Lamarche, F.: Proof nets for intuitionistic linear logic: Essential nets. Technical Report 00347336, INRIA (2008)
12. Roorda, D.: Resource logics: proof-theoretical investigations. PhD thesis, Universiteit van Amsterdam (1991)
13. de Groote, P.: An algebraic correctness criterion for intuitionistic multiplicative proof-nets. Theoretical Computer Science 224(1-2), 115–134 (1999)

Characterizing Discontinuity in Constituent Treebanks

Wolfgang Maier and Timm Lichte*

University of Tübingen
{wo.maier,timm.lichte}@uni-tuebingen.de

Abstract. Measures for the degree of non-projectivity of dependency grammar have received attention both on the formal and on the empirical side. The empirical characterization of discontinuity in constituent treebanks annotated with crossing branches has nevertheless been neglected so far. In this paper, we present two measures for the characterization of both the discontinuity of constituent structures and the non-projectivity of dependency structures. An empirical evaluation on German data as well as an investigation of the relation between the measures and grammars extracted from treebanks shows their relevance.

1 Introduction

Discontinuous phrases are common in natural language. They occur particularly frequently in languages with a relatively free word order like German, but also in fixed word order languages, like English or Chinese. Treebank annotation must account for sentences containing such phrases. In constituent treebanks, however, a direct annotation of discontinuities which would require crossing branches is generally not allowed. Instead, an annotation backbone based on context-free grammar is extended by a labeling mechanism which accounts for discontinuous phrases, sometimes in combination with traces. Examples for such treebanks include the Penn Treebank (PTB) [1], the Spanish 3LB [2], and the German TüBa-D/Z [3].

The German NeGra [4] and TIGER [5] treebanks are among the few notable exceptions to this kind of annotation. In both treebanks, crossing branches are allowed and discontinuous constituents are annotated directly. As an illustration, Fig. 1 shows the constituent annotation of (1), a sentence from NeGra involving two discontinuous VPs. The discontinuity is due to the leftward extraposition of the PP object *Darüber*.

(1) Darüber muß nachgedacht werden.
     Thereof must thought   be.
     "Thereof must be thought."

---

* Thanks to Laura Kallmeyer, Gerhard Jäger, Armin Buch, Marco Kuhlmann and the three anonymous reviewers for helpful comments and suggestions.

P. de Groote, M. Egg, and L. Kallmeyer (Eds.): FG 2009, LNAI 5591, pp. 167–182, 2011.

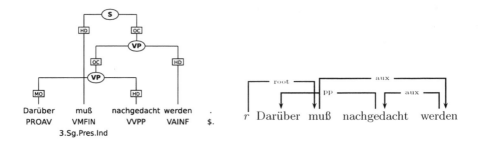

Fig. 1. Constituent and dependency analysis for a NeGra sentence

Measures for the degree of discontinuity of constituent trees with crossing branches in these treebanks have not been subject to research, but warrant an investigation. First, linguistic phenomena like long-distance dependencies, which give rise to discontinuous structures, are often assumed to adhere to certain restrictions. Furthermore, in application contexts like statistical parsing, where treebank trees are interpreted as derivation structures of a grammar formalism, the information encoded by the crossing branches is simply discarded in order to avoid processing difficulties: Commonly, the trees are transformed into non-crossing structures and a (probabilistic) context-free grammar is read off the resulting trees [6,7].

The situation for dependency grammar is different. Dependency treebanks generally contain *non-projective* graphs which, in contrast to *projective* graphs, contain nodes with an discontinuous yield (see Fig. 1 for a dependency analysis of (1)). In recent years, motivated by processing concerns, particularly the formal characterization of non-projectivity has received attention. See [8] for a survey. Among other measures, *gap degree* [9] and *well-nestedness* [10] have emerged. Both of them were shown to empirically describe non-projective dependency data well. They can furthermore be exploited for parsing [11].

The contribution of this paper is to give a characterization of gap degree and well-nestedness of constituent trees with the same descriptive and practical relevance as the corresponding concepts in dependency grammar. For this purpose, we show that both gap degree and well-nestedness can be formulated as equivalent properties of different kinds of graphs, more precisely, of dependency graphs and constituent trees. We then explore the relation between the measures and grammars extracted from dependency and constituent treebanks and conduct an empirical investigation on German data that shows the practical relevance of the measures.

In the following section, we introduce new definitions of gap degree and well-nestedness. In Sect. 3 we investigate relation between the measures and grammars extracted from treebanks. In Sect. 4, we empirically review the measures for non-projectivity of dependency graphs and discontinuity of constituent trees on German data. Sect. 5 summarizes the article and presents future work.

2   Discontinuity Measures

In the following, we define both dependency structures and constituent structures in terms of trees, followed by the definitions of gap degree and well-nestedness.

2.1   Syntactic Structures

Definition 1 (Tree). *Let $D = (V, E, r)$ be a directed graph with V a set of nodes, $E : V \times V$ a set of arcs and $r \in V$ the root node. D is a* labeled tree *iff*

1. *D is acyclic and connected,*
2. *all nodes $V \setminus \{r\}$ have in-degree 1, r has in-degree 0,*
3. *D disposes of two disjoint alphabets L_V, L_E of node and edge labels and the corresponding labeling functions $l_V : V \to L_V$ and $l_E : E \to L_E$.*

The nodes with out-degree 0 are called *leaves*. We write $v_1 \to v_2$ for $\langle v_1, v_2 \rangle \in E$. $\stackrel{*}{\to}$ is the reflexive transitive closure of E. We say that v_1 *dominates* v_2 iff $v_1 \stackrel{*}{\to} v_2$.

On the basis of trees, we define *dependency structures* and *constituent structures*.

Definition 2 (Dependency structure[1]). *A dependency structure for a sentence $s = w_1 \cdots w_n$ is a labeled tree $D_{dep} = (V, E, r)$ with the labeling functions l_V and l_E, such that*

1. *$l_V : V \to \{0, \ldots, n\}$, $l_V(v) \neq 0$ for all $v \in V \setminus \{r\}$, $l_V(r) = 0$,*
2. *$l_E : E \to L_E$, L_E being a set of dependency edge labels.*

Note that the root node in dependency structures is mapped to 0, which is no terminal position and is therefore not included in its yield (see below).

Definition 3 (Constituent structure). *A constituent structure for a sentence $s = w_1 \cdots w_n$ is a labeled tree $D_{con} = (V, E, r)$ with the labeling functions l_V and l_E, such that*

1. *$l_V : V \to \{1, \ldots, n\} \cup N$, N being a set of syntactic category labels disjoint from $\{1, \ldots, n\}$, such that for a $v \in V$, $l_V(v) \in \{1, \ldots n\}$ if v is a leaf and $l_V(v) \in N$ otherwise,*
2. *$l_E : E \to L_E$, L_E being a set of grammatical function labels.*

Definition 4 (Syntactic structure). *We call $D = (V, E, r)$ a syntactic structure iff it is either a dependency structure or a constituent structure. The* yield *π_v of a $v \in V$ is the set $\{i \in \mathbb{N}^+ \mid$ there is a $v' \in V$ such that $v \stackrel{*}{\to} v'$ and $l_V(v') = i\}$.*

---

[1] Our definition follows the definition of *Abhängigkeitsbäumen* (dependency trees) in [12].

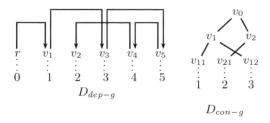

Fig. 2. Gap degree in syntactic structures

2.2 Discontinuity Measures

Both gap degree and well-nestedness can be defined on the yields of syntactic structures. Intuitively, a gap is a discontinuity in the yield of a node and the gap degree is a measure for the amount of discontinuity in syntactic structures.

Definition 5 (Gap degree). *Let $D_{syn} = (V, E, r)$ be a syntactic structure.*

1. *Let π_v be the yield of a $v \in V$. A gap is a pair (i_n, i_m) with $i_n, i_m \in \pi_v$ and $i_n + 2 \leq i_m$ such that there is no $i_k \in \pi_v$ with $i_n < i_k < i_m$. The gap degree d of a node v in a syntactic structure is the number of gaps in π_v.*
2. *The gap degree d of a syntactic structure D_{syn} is the maximal gap degree of any of its nodes.*

A dependency structure with gap degree > 0 is called *non-projective*; a constituent structure with gap degree > 0 is called *discontinuous*.

In other words, the gap degree corresponds to the maximal number of times the yield of a node is interrupted. n gaps of a node entail $n + 1$ uninterrupted yield intervals. This is reflected in the measure of *block degree* [8, pp. 35], which is in fact the gap degree plus 1. Fig. 2 shows an example. Both syntactic structures in D_{dep-g} and D_{con-g} have gap degree 1. In D_{dep-g} both yields π_{v_4} and π_{v_5} have the maximal gap degree 1 due to the fact that they do not contain 3. In D_{con-g}, v_1 has gap degree 1 because 2 is not included in its yield. D_{dep} and D_{con} in Fig. 3 both have gap degree 0.

We now define *well-nestedness* (as opposed to *ill-nestedness*) as another property of syntactic structures. Intuitively, in a well-nested structure, it holds for all nodes which do not stand in a dominance relation that their yields do not interleave.

Definition 6 (Well-Nestedness[2]). *Let $D_{syn} = (V, E, r)$ be a syntactic structure. D_{syn} is well-nested iff there are no disjoint yields π_{v_1}, π_{v_2} of nodes $v_1, v_2 \in V$ such that for some $i_1, i_2 \in \pi_{v_1}$ and $j_1, j_2 \in \pi_{v_2}$ it holds that $i_1 < j_1 < i_2 < j_2$.*

It is easy to see that an ill-nested syntactic structure must necessarily have a gap degree ≥ 1. Fig. 3 shows well-nested and ill-nested constituent structures

---

[2] If the considered yields are not restricted to be disjoint, the syntactic structures are called *planar* or *weak non-projective* (see [13,14]).

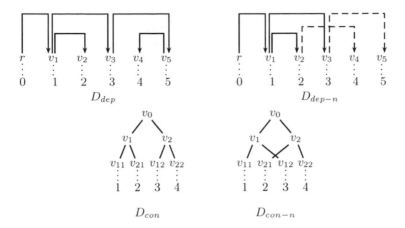

Fig. 3. Well-nestedness and ill-nestedness in syntactic structures

and dependency structures. D_{con} and D_{dep} are well-nested, while D_{con-n} and D_{dep-n} are not. Note that, while D_{con} and D_{dep} have gap degree 0 in addition to being well-nested, both D_{con-g} and D_{dep-g} in Fig. 2 are well-nested but have a gap degree greater than 1.

Gap degree and well-nestedness had not been defined for constituent trees before. However, with respect to dependency structures our definitions do correspond to the definitions of gap degree and well-nestedness from the literature [9,10,14].

The definition of well-nestedness (resp. ill-nestedness) does not provide a notion of a degree. For this purpose, to our knowledge, two measure have been introduced so far in the context of dependency structures. [13] introduces *level types* and [15] introduce *strongly ill-nested structures*. While both measures do characterize well the data considered in the resp. articles, the first measure is path-based rather than yield-based and the second one only discriminates very complex dependency structures which are unlikely to occur in a natural language context. In the following, we define a new measure called *k-ill-nestedness* which we intend to intuitively capture the degree of interleaving of yields in constituent structures and dependency structures. k stands for the number of disjoint yields that interleave with some other single yield which is disjoint from them.

Definition 7 (k-Ill-Nestedness). *Let $D_{syn} = (V, E, r)$ be a syntactic structure.*

1. *The degree k of ill-nestedness of a node $v \in V$ is the number k of nodes $v_1, \ldots, v_k \in V$ with the yields $\pi_{v_1}, \ldots, \pi_{v_k}$ such that for some $i_1, i_2 \in \pi_v$ and $j_1^{(m)}, j_2^{(m)} \in \pi_m$, $1 \le m \le k$, it holds that $i_1 < j_1^{(m)} < i_2 < j_2^{(m)}$ or $j_1^{(m)} < i_1 < j_2^{(m)} < i_2$.*
2. *D_{syn} is said to be k-ill-nested iff k is the maximal degree of ill-nestedness of any of its nodes.*

Note that well-nestedness as defined above is equivalent to 0-ill-nestedness. The dependency structure D_{dep-n} and the constituent structure D_{con-n} in Fig. 3 are 1-ill-nested. An example for 2-ill-nested syntactic structures is D_{con-lr} in Fig. 4.

2.3   Further Properties of Constituent Structures

We now give a further formal characterization of constituent structures. We first define the *maximal nodes* of a gap. Informally speaking, the entire yield of a maximal node lies in the gap, but the yield of its parent node does not.

Definition 8 (Maximal node). *Let* $D_{con} = (V, E, r)$ *be a constituent structure,* π_v *the yield of a* $v \in V$ *and* (i_1, i_2) *a gap in* π_v. *Then* $v_{max} \in V$ *is a maximal node of* (i_1, i_2) *iff*

1. *for all* $j \in \pi_{v_{max}}$, *it holds that* $i_1 < j < i_2$,
2. *there is a node* $u \in V$ *with the yield* π_u *such that* $u \to v_{max}$, *and there is a* $k \in \pi_u$ *with* $k \leq i_1$ *or* $k \geq i_2$.

A gap (i_1, i_2) can have more than one maximal node, and the combined yield of all maximal nodes is the set $\{i \mid i_1 < i < i_2\}$. As an example, consider D_{con-n} in Fig. 3. The only maximal node of the gap $(1, 3)$ of v_1 is v_{21}.

Intuitively, in a well-nested constituent structure $D_{con} = (V, E, r)$, all gaps are filled from "above". That means that all gap fillers (i.e., the parent nodes of all maximal nodes) of all gaps (i_1, i_2) in the yield of a $v \in V$ immediately dominate v itself. As an example, compare the well-nested structure D_{con-g} (Fig. 2) with the ill-nested structure D_{con-n} (Fig. 3): In the first, v_0 is the maximal node of the gap of v_1 and $v_0 \to v_1$, while in D_{con-n}, v_2 is the maximal node of the gap of v_1 and $v_2 \not\to v_1$. We formalize this intuition in Lemma 1.

Lemma 1. *Let* $D_{con} = (V, E, r)$ *be a well-nested constituent structure,* $v \in V$ *a node with gap degree* ≥ 1, (i_1, i_2) *a gap in* π_v *and* v_{max} *a maximal node of* (i_1, i_2). *There is no node* v' *with* $v' \to v_{max}$ *and* $v' \not\to v$.

Proof. By contradiction. Assume that there is a node v' with $v' \to v_{max}$ and $v' \not\to v$. According to the definition of maximal nodes, there must be a $k_i \in \pi_{v'}$ with $k_i < i_1$ or $k_i > i_2$ and a k_j with $i_1 < k_j < i_2$. That means that it is either the case that $k_i < i_1 < k_j < i_2$ or $i_1 < k_j < i_2 < k_i$, both of which contradict the definition of well-nestedness.   □

Ill-nested constituent structures are thus constituent structures in which some gap is filled from a direction other than "above", in other words, in which the gap filler does not dominate the node with the gap. Gaps in such structures can intuitively be filled from the left, from the right, or from both sides. If we assume an implicit ordering of the nodes based on the smallest number in their yield, this intuition holds even though we deal with unordered trees. Such an ordering is in fact the basis for all visual representations of constituent structures in this paper. Note that if a gap is filled from the left or the right and additionally from above, we are still dealing with an ill-nested structure (cf. Lemma 1).

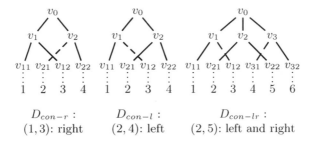

$$D_{con-r}: \qquad D_{con-l}: \qquad D_{con-lr}:$$
$$(1,3): \text{right} \qquad (2,4): \text{left} \qquad (2,5): \text{left and right}$$

Fig. 4. Gap filler position in ill-nested constituent structures

Definition 9 (Gap filler locations). *Let $D_{con} = (V, E, r)$ be an ill-nested constituent structure, $v \in V$ a node with gap degree ≥ 1, (i_1, i_2) a gap in π_v and v_{max} a maximal node of (i_1, i_2). We say that*

1. *(i_1, i_2) is filled from the left iff there is a node v' with the yield $\pi_{v'}$ and $v' \to v_{max}$ such that there is an $i_l \in \pi_{v'}$ with $i_l < i_1$ and*
2. *(i_1, i_2) is filled from the right iff there is a node v' with the yield $\pi_{v'}$ and $v' \to v_{max}$ such that there is an $i_l \in \pi_{v'}$ with $i_l > i_1$.*

Fig. 4 shows three example trees with different gap filler locations.

3   Properties of Extracted Grammars

Both non-projective dependency structures and discontinuous constituent structures, as occurring in treebanks, can be interpreted as derivation structures of linear context-free rewriting systems (LCFRS) [16], resp. of the equivalent formalism of simple range concatenation grammar (sRCG) [17]. [18] present a grammar extraction algorithm for constituent structures and [11] presents an extraction algorithm for dependency structures. In the following we resume previous work by presenting simple RCG together with an extraction algorithm for this formalism on syntactic structures.

3.1   Simple Range Concatenation Grammar

Definition 10 (Ordered simple k-RCG [17]). *A simple RCG is a tuple $G = (N, T, V, P, S)$ where N is a finite set of predicate names with an arity function dim: $N \to \mathbb{N}$, T and V are disjoint finite sets of terminals and variables, P is a finite set of clauses of the form*

$$A(\alpha_1, \ldots, \alpha_{dim(A)}) \to A_1(X_1^{(1)}, \ldots, X_{dim(A_1)}^{(1)}) \cdots A_m(X_1^{(m)}, \ldots, X_{dim(A_m)}^{(m)})$$

for $m \geq 0$ where $A, A_1, \ldots, A_m \in N$, $X_j^{(i)} \in V$ for $1 \leq i \leq m, 1 \leq j \leq dim(A_i)$ and $\alpha_i \in (T \cup V)^*$ for $1 \leq i \leq dim(A)$, and $S \in N$ is the start predicate name with $dim(S) = 1$. For all $p \in P$, it holds that every variable X occurring in c occurs exactly once in the left-hand side (LHS) and exactly once in the RHS.

*A simple RCG is ordered if for all $p \in P$, it holds that if a variable X_1
precedes a variable X_2 in some RHS predicate, then X_1 also precedes X_2 in
the LHS predicate. A simple RCG $G = (N, T, V, P, S)$ is a k-RCG if for all
$A \in N, dim(A) \leq k$. k is called the* fan-out *of G. The* rank *of G is the maximal
number of predicates on the RHS of one of its clauses.*

For the definition of the language of a simple RCG, we borrow the LCFRS
definitions.

Definition 11 (Language of a simple RCG). *Let $G = \langle N, T, V, P, S \rangle$ be a
simple RCG. For every $A \in N$, we define the yield of A, $yield(A)$ as follows:*

1. *For every $A(\boldsymbol{\alpha}) \to \varepsilon$, $\boldsymbol{\alpha} \in yield(A)$;*
2. *For every clause*

$$A(\alpha_1, \ldots, \alpha_{dim(A)}) \to A_1(X_1^{(1)}, \ldots, X_{dim(A_1)}^{(1)}) \cdots A_m(X_1^{(m)}, \ldots, X_{dim(A_m)}^{(m)})$$

   *and all $\boldsymbol{\tau_i} \in yield(A_i)$ for $1 \leq i \leq m$, $\langle f(\alpha_1), \ldots, f(\alpha_{dim(A)}) \rangle \in yield(A)$
   where f is defined as follows:*
   (a) $f(t) = t$ for all $t \in T$,
   (b) $f(X_j^{(i)}) = \boldsymbol{\tau_i}(j)$ for all $1 \leq i \leq m, 1 \leq j \leq dim(A_i)$,
   (c) $f(xy) = f(x)f(y)$ for all $x, y \in (T \cup V)^+$ and
   (d) $f(\varepsilon) = \varepsilon$.
3. *Nothing else is in $yield(A)$.*

The language is then $\{w \,|\, \langle w \rangle \in yield(S)\}$.

3.2   Grammar Extraction

We can combine both of the afore-cited grammar extraction algorithms into a
single algorithm. We extract a simple RCG G from a syntactic structure $D =
(V, E, r)$ over a sentence $w_1 \ldots w_n$ as follows. For each $v_0 \in V$ with the children
$v_1, \ldots, v_k$, $k \geq 0$, we construct a clause $p = \psi_0 \to \psi_1 \cdots \psi_k$, where $\psi_i, 0 \leq i \leq k$,
are predicates which we construct as follows. If D is a dependency structure,
then ψ_0 receives the label of the incoming arc of v_0 and $\psi_i, 1 \leq i \leq k$, receives
the label of the arc between v_0 and v_i; otherwise, $\psi_i, 0 \leq i \leq k$, receives the node
label of v_i. To the name of each $\psi_i, 0 \leq i \leq k$, we append the block degree of v_i.
In order to determine the predicate arguments, for each $u \in \pi_0$, we introduce a
variable X_u. Then for all $\psi_i, 0 \leq i \leq k$, the following conditions must hold.

1. The concatenation of all arguments of ψ_i is the concatenation of all $\{X_u \mid
   u \in \pi_i\}$ such that for all $p, q \in \pi_i, p < q$, it holds that $X_p \prec X_q$.
2. A variable X_i with $1 \leq i < n$ is the right boundary of an argument of ψ_i iff
   $i + 1 \notin \pi_i$, i.e., and argument boundary is introduced at each gap.

If D is a dependency structure and $l_V(v_0) = i$, we exchange the variable X_i in
ψ_0 with w_i. If D is a constituent structure, $l_V(v_0) = i$ and $l_V(v') = t_i, v' \to v_0$,

$$S1(X_1X_2X_3) \rightarrow VP2(X_1,X_3) \; VMFIN1(X_2)$$
$$VP2(X_1, X_2X_3) \rightarrow VP2(X_1,X_2) \; VAINF1(X_3)$$
$$VP2(X_1, X_2) \rightarrow PROAV1(X_1) \; VVPP1(X_2)$$
$$PROAV(Dar\ddot{u}ber) \rightarrow \varepsilon, \; VMFIN(mu\ss) \rightarrow \varepsilon$$
$$VVPP(nachgedacht) \rightarrow \varepsilon, \; VAINF(werden) \rightarrow \varepsilon.$$

$$pp(Dar\ddot{u}ber) \rightarrow \varepsilon$$
$$root(X_1 mu\ss X_3) \rightarrow aux(X_1,X_3)$$
$$aux(X_1, nachgedacht) \rightarrow pp(X_1)$$
$$aux(X_1, X_2 werden) \rightarrow aux(X_1,X_2)$$

Fig. 5. Simple RCGs obtained from constituent structures (left) and from dependency structures (right)

we add a so-called *lexical clause* $t_i(w_i) \rightarrow \varepsilon$ to the grammar, t_i being the part-of-speech tag of w_i. Sequences of more than one variable which are a RHS argument are collapsed into a single variable both on the LHS and the RHS of the clause. The completed clause p is added to the grammar.

As an example, Fig. 5 shows the two grammars extracted from the constituent and dependency structures in Fig. 1.

3.3   Grammar Properties

The gap degree and the ill-nestedness of the syntactic structure used for grammar extraction determine certain properties of the extracted grammar. The properties in Corollary 1 are immediately obvious.

Corollary 1 (Grammar Properties). *Let G be a simple RCG of rank r and fan-out g. A syntactic structure D derived from G has at most gap degree $g-1$ and does not have nodes with more than r children.*

Assuming the relation between simple RCGs and syntactic structures given by the grammar extraction algorithm above, we can also draw conclusions from the clauses of a simple RCG G on the well-nestedness of its derivations. Lemma 2 relates the interleaving of the variables in the arguments of clauses with the conditions given by the definition of well-nestedness. For simplicity, we assume that all clauses $p \in P$ contain only continuously numbered variables X_1 through X_m in their LHS predicate. Furthermore, we introduce the function $\eta : P \times \mathbb{N} \rightarrow \mathbb{N}^+$, which returns the numbers of the variables used in the arguments of the ith RHS predicate of a clause.

Lemma 2 (Ill-Nestedness and Grammars). *Let G be a ordered simple RCG. If it produces k-ill-nested derivations with $k \geq 1$, then there is a clause $p \in P$, $p = \psi_0 \rightarrow \psi_1 \cdots \psi_m$ such that there is a pair of predicates (ψ_i, ψ_j), $1 \leq i < j \leq m$ for which it holds that $i_1, i_2 \in \eta(p, i)$ and $j_1, j_2 \in \eta(p, j)$ with $i_1 < j_1 < i_2 < j_2$.*

Proof. By contradiction. Assume there is no such clause and G produces ill-nested derivations. Due to the definition of well-nestedness, there must be a derivation $D = (V, E, r)$ of G with disjoint yields π_{v_1}, π_{v_2} of nodes $v_1, v_2 \in V$ such that for some $i_1, i_2 \in \pi_{v_1}$ and $j_1, j_2 \in \pi_{v_2}$ it holds that $i_1 < j_1 < i_2 < j_2$. The fact that π_{v_1} and π_{v_2} are disjoint entails that v_1 and v_2 do not dominate each other. Furthermore, it entails that the yield π_{lca} of their least common ancestor v_{lca} must contain both π_{v_1} and π_{v_2} and that for all yields π' of nodes v' with $v_{lca} \xrightarrow{*} v'$, it holds that $\pi' \cap \pi_{v_1} = \emptyset$ or $\pi' \cap \pi_{v_2} = \emptyset$. Assume v_{lca}

and its children have been generated by a clause $\psi_{lca} \to \psi_{lca}^1 \cdots \psi_{lca}^m$. Due to the aforementioned condition on the yields of all nodes dominated by v_{lca} and condition 1 in the extraction algorithm, there must be at least two predicates $\psi_{lca}^i, \psi_{lca}^j, 1 \le i < j \le m$ for which it holds that $i_1, i_2 \in \eta(p, i)$ and $j_1, j_2 \in \eta(p, j)$ with $i_1 < j_1 < i_2 < j_2$. This contradicts our initial assumption.   □

Note that given a simple RCG which produces k-illnested structures with $k \ge 1$, it is not possible to determine the exact k without inspecting all possible derivations, i.e., the k can not be determined by properties of the clauses of the grammar alone. This is due to the fact that k-illnestedness rises through the interplay of more than one clause. See furthermore [19], for a grammar definition where the well-nestedness property is included.

4   Empirical Investigation

4.1   Non-projective Dependency Structures

The empirical relevance of gap degree and well-nestedness has been shown for dependency treebanks in works such as [14] on the basis of the Prague Dependency Treebank (PDT) [20] and the Danish Dependency Treebank (DDT) [21]. We have extended these previous investigations on two other treebanks, namely the dependency versions of the NeGra and TIGER, which we will call TIGER-Dep and NeGra-Dep. Both have been built with the dependency converter for TIGER-style trees from [22]. We are aware of the fact that all dependency conversion methods introduce undesired noise, but we choose this well-established method rather than using other German dependency data sets like the CoNLL-X TIGER data (used by [13]) or the very small TIGER-DP [23] due to our desire of obtaining comparable dependency structures for both constituent treebanks.

Since punctuation is generally not attached to the trees and therefore not part of the annotation, it has been removed in a preprocessing step from all trees in both treebanks prior to the conversion. This lead to the exclusion of a handful of sentences from TIGER and NeGra, since they consisted only of punctuation. Tab. 1 contains the gap degree figures and ratios of well-nestedness of the PDT and the DDT, borrowed from [14], and our findings for NeGra-Dep and TIGER-Dep.

The gap degree figures of NeGra-Dep and TIGER-Dep lie in the same range as the figures of DDT and PDT. A closer look at the dependency annotations of TIGER-Dep and NeGra-Dep reveals that the most common causes for a high gap-degree (≥ 3) are enumerations, appositions and parenthetical constructions.

As [8, p. 62] remarks, the well-nestedness constraint seems to be a good extension of projectivity since almost all dependency structures found in treebanks adhere to it. This is confirmed by our quantitative findings in the German treebanks. With the exception of a single 2-ill-nested dependency structure[3]

---

[3] The edge which is responsible for the 2-ill-nestedness cannot be linguistically motivated and is an artefact introduced by the conversion producedure.

Table 1. Gap degree and well-nestedness in DDT and PDT (figures from [14]) and in NeGra-Dep and TIGER-Dep

	DDT		PDT		NeGra-Dep		TIGER-Dep	
number of sent.	4,393		73,088		20,597		40,013	
av. sent. length	18		17		15		16	
gap degree 0	3,732	84.95%	56,168	76.85%	16,695	81.06%	32,079	80.17%
gap degree 1	654	14.89%	16,608	22.72%	3,662	17.78%	7,466	18.66%
gap degree 2	7	0.16%	307	0.42%	225	1.09%	438	1.09%
gap degree 3	–	–	4	0.01%	12	0.05%	22	0.05%
gap degree 4	–	–	1	< 0.01%	2	0.01%	4	0.01%
gap degree ≥5	–	–	–	–	1	< 0.01%	4	0.01%
well-nested	4,388	99.89%	73,010	99.89%	20,472	99.39%	39,750	99.34%
1-ill-nested	?		?		124	0.60%	263	0.66%
2-ill-nested	?		?		1	< 0.01%	–	–

in NeGra-Dep, the structures which do not adhere to the well-nestedness constraint are all 1-ill-nested. Note that 1-ill-nestedness is a stronger constraint than non-well-nestedness. A closer linguistic inspection of the ill-nested dependency structures in Negra-Dep and TIGER-Dep shows that most common reason for ill-nestedness is not erroneous annotation etc., but linguistically acceptable analyses of extraposition phenomena. Fig. 6 shows a typical ill-nested dependency analysis of a sentence of NeGra. The edges relevant for the ill-nestedness are dashed. In this sentence, the subject noun *ein Modus* dominates a non-adjacent, extraposed relative clause, while being surrounded by a disjoint subtree, namely the non-finite main verb *eingespielt* and its dependent. The annotation can be

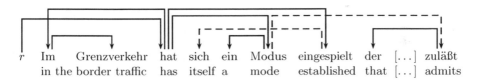

Fig. 6. An ill-nested dependency structure in Negra-Dep

related to linguistic generalizations on dependency structures discussed in the literature (particularly on German):

1. The subject depends on the finite verb [12, p. 110][24, pp. 83].
2. The non-finite verb depends on the finite verb and governs its objects and modifying expressions[25, p. 189]. In our treebanks, this is only true for objects and modifying expressions outside the *Vorfeld* since *Vorfeld* material is systematically attached to the finite verb by the conversion procedure.
3. Extraposed material is dependent on its antecedent [12, pp. 130][24, pp. 101].

In the remaining sentences, there are two phenomena which give rise to ill-nested structures, namely coordinated structures and discontinuous subjects. Fig. 7

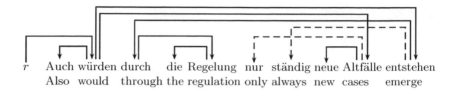

r   Auch würden durch   die Regelung nur ständig neue Altfälle entstehen
     Also would   through the regulation only always new cases   emerge

Fig. 7. Ill-nestedness due to a discontinuous subject

shows the annotation of (2), an example from TIGER-Dep for a discontinuous subject. Again, the relevant edges are dashed.

(2)   Auch würden durch   die Regelung nur ständig neue Altfälle entstehen.
       Also would   through the regulation only always new cases   emerge
       "Another effect of the regulation would be constantly emerging new cases."

The ill-nested annotation of the coordination cases is largely disputable; however, this can be explained with the lack of a general linguistic theory of coordination [26]. The situation for discontinuous subjects is clearer, since one can argue that the components of the discontinuous subject distinguish themselves from the material in the gap by making up a semantic unit.

To sum up, ill-nested dependency annotation in NeGra-Dep and TIGER-Dep can generally be linguistically justified and is not due to annotation oddities. An accurate linguistic survey of (k)-ill-nestedness of structures in DDT and PDT has not been presented in the literature and is left for future work.

4.2   Discontinuous Constituent Structures

In the following, we investigate gap degree and well-nestedness of constituent treebanks in order to verify if both measures are as informative for constituent structures as they are for dependency structures. We conduct our study on the constituent versions of the treebanks in the previous section, using exactly the same set of sentences (with removed punctuation). The results are summarized in Fig. 2.

The constituent gap degree figure of both German treebanks again lie together. We found that the number of constituent structures with gap degree ≥ 3 is considerably lower than the corresponding number of dependency structures. The reason is that the phenomena which cause a high gap degree in dependency structures (enumerations and appositions) generally receive a constituent structure without gaps. The most frequent reasons for gaps in constituent structures are parenthetical constructions, as well as finite verbs, subjects and negative markers, which are generally annotated as immediate constituents of the highest S node and therefore may cause gaps in the VP yield.

Tab. 2 shows that the ratio of ill-nested structures in constituent data is comparable to the ratio in dependency data. This suggests that ill-nestedness has a comparable explanatory value as a constraining feature for constituent structures. The only degree of ill-nestedness that can be observed is 1-ill-nestedness.

Table 2. Constituent gap degree of TIGER/NeGra trees

	NeGra	TIGER
total	20597	40013
gap degree 0	14,648 72.44%	28,414 71.01%
gap degree 1	5,253 24.23%	10,310 25.77%
gap degree 2	687 3.30%	1,274 3.18%
gap degree 3	9 0.04%	15 0.04%
well-nested	20,339 98.75%	39,573 98.90%
1-ill-nested	258 1.25%	440 1.10%
2-ill-nested	– –	– –

A linguistic inspection of the ill-nested constituent structures shows that most of them are ill-nested due to the interplay of several annotation principles. Again, most of the ill-nested constituent structures in TIGER and NeGra arise from extraposition phenomena. Furthermore we can also find cases of discontinuous subjects annotated with ill-nested structures. Other than for the dependency structures, coordination is no trigger for ill-nestedness in the constituent data.

As an example for ill-nested constituent structure, see the embedded sentence (3) and its tree annotation in Fig. 8.

(3)   . . . ob   auf deren Gelände der Typ von Abstellanlage   gebaut
        . . . whether on their   premises the type of   parking facility built
        werden könne, der   . . .
        be   could, which . . .
        "whether on their premises precisely the type of parking facility could be
        built, which . . . "

The two overlapping, disjunctive constituents are the lower VP, and the NP with its extraposed relative clause.

The following underlying annotation principles seem to be respected throughout:

1. the subject is an immediate constituent of the sentence;
2. the finite verb is another immediate constituent (and the head) of the sentence;
3. the non-finite verb is the head of another immediate constituent that also includes objects and modifying expressions;
4. extraposed material is included in the antecedent constituent.

We will not argue in favor or against these annotation principles from a linguistic point of view. A mapping to common linguistic theories is highly non-trivial, since very different means of expressing constituency relations and a variety of shapings of constituent structures would have to be taken into account. On the other hand, the similarity to the above stated annotation principles for dependency structures is striking.

Ill-nestedness does not affect the same set of structures across treebank variants, i.e., the ill-nested dependency structures are no subset of the ill-nested

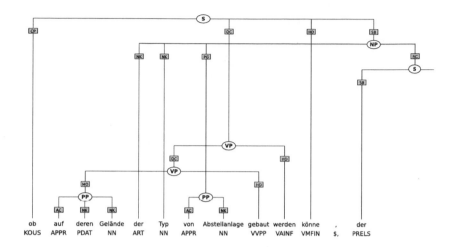

Fig. 8. An ill-nested sentence from NeGra

constituent structures or vice versa. We leave for future work an exhaustive investigation of the differences.

5   Conclusion

We have presented a formal account for two measures of discontinuity, resp. non-projectivity, namely gap degree and well-nestedness. Our definitions emphasize that the notions of gap degree and well-nestedness are independent of the type of syntactic structures considered, since they apply to both dependency and constituent structures. Concerning their application on dependency structures, they correspond to the well-known definitions from the literature. We have conducted an empirical study on two different versions of two treebanks, one annotated with constituents with crossing branches, the other one annotated with non-projective dependencies. The results show that the explanatory value of the measures applied on constituent trees is as high as for dependency structures. Even though well-nestedness has an almost perfect coverage on both constituent and dependency data, a linguistic inspection of ill-nested dependency and constituent structures shows that there exist linguistic phenomena for which ill-nestedness can indeed be linguistically justified.

As mentioned before, dependency conversion procedures introduce undesired noise. Therefore, in future work, we plan to undertake an exhaustive investigation of the exact effects of the conversion on critical examples. Furthermore, we plan to conduct a study on the Bulgarian BulTreebank, which contains both discontinuous constituent and non-projective dependency annotation [27], and to investigate the possibilities of introducing crossing edges in constituent treebanks which have an annotation backbone based on context-free grammar (e.g. the Penn Treebank).

References

1. Marcus, M.P., Santorini, B., Marcinkiewicz, M.A.: Building a large annotated corpus of English: The Penn Treebank. Computational Linguistics 19(2), 313–330 (1994)
2. Civit, M., Martí Antònín, M.A.: Design principles for a Spanish treebank. In: Proceedings of the 1st Workshop on Treebanks and Linguistic Theories, Sozopol, Bulgaria (2002)
3. Telljohann, H., Hinrichs, E., Kübler, S., Zinsmeister, H.: Stylebook for the Tübingen Treebank of Written German (TüBa-D/Z). Technischer Bericht, Seminar für Sprachwissenschaft, Universität Tübingen, Tübingen (July 2006) Revidierte Fassung
4. Skut, W., Krenn, B., Brants, T., Uszkoreit, H.: An annotation scheme for free word order languages. In: Proceedings of the 5th Applied Natural Language Processing Conference, Washington, DC, pp. 88–95 (1997)
5. Brants, S., Dipper, S., Hansen, S., Lezius, W., Smith, G.: The TIGER Treebank. In: Proceedings of the 1st Workshop on Treebanks and Linguistic Theories, Sozopol, Bulgaria, pp. 24–42 (2002)
6. Kübler, S., Hinrichs, E.W., Maier, W.: Is it really that difficult to parse German? In: Proceedings of the 2006 Conference on Empirical Methods in Natural Language Processing, Sydney, Australia, pp. 111–119 (July 2006)
7. Boyd, A.: Discontinuity revisited: An improved conversion to context-free representations. In: Proceedings of the 45th Annual Meeting of the Association of Computational Linguistics, the Linguistic Annotation Workshop, Prague, Czech Republic, pp. 41–44 (2007)
8. Kuhlmann, M.: Dependency Structures and Lexicalized Grammars. PhD thesis, Saarland University (2007)
9. Holan, T.: Kuboň, V., Oliva, K., Plátek, M.: Two useful measures of word order complexity. In: Workshop on Processing of Dependency-Based Grammars, Montréal, Canada, pp. 21–29 (1998)
10. Bodirsky, M., Kuhlmann, M., Möhl, M.: Well-nested drawings as models of syntactic structure. In: Proceedings of the 10th Conference on Formal Grammar and the 9th Meeting on Mathematics of Language (FG-MOL 2005), Edinburgh, UK (2005)
11. Kuhlmann, M., Satta, G.: Treebank grammar techniques for non-projective dependency parsing. In: Proceedings of the 12th Conference of the European Chapter of the Association for Computational Linguistics, Athens, Greece (2009)
12. Kunze, J.: Abhängigkeitsgrammatik. Studia grammatica, vol. 12. Akademie-Verlag, Berlin (1975)
13. Havelka, J.: Beyond projectivity: Multilingual evaluation of constraints and measures on non-projective structures. In: Proceedings of the 45th Annual Meeting of the Association of Computational Linguistics, pp. 608–615 (2007)
14. Kuhlmann, M., Nivre, J.: Mildly non-projective dependency structures. In: Proceedings of the COLING/ACL 2006 Main Conference Poster Sessions, Sydney, Australia (2006)
15. Gómez-Rodríguez, C., Weir, D., Carroll, J.: Parsing mildly non-projective dependency structures. In: Proceedings of the 12th Conference of the European Chapter of the ACL (EACL 2009), Athens, Greece, pp. 291–299. Association for Computational Linguistics (March 2009)

16. Vijay-Shanker, K., Weir, D., Joshi, A.: Characterising structural descriptions used by various formalisms. In: Proceedings of ACL (1987)
17. Boullier, P.: Proposal for a natural language processing syntactic backbone. Rapport de Recherche RR-3342, Institut National de Recherche en Informatique et en Automatique, Le Chesnay, France (1998)
18. Maier, W., Søgaard, A.: Treebanks and mild context-sensitivity. In: Proceedings of the 13th Conference on Formal Grammar 2008, Hamburg, Germany, pp. 61–76 (2008)
19. Kracht, M.: The Mathematics of Language. Mouton de Gruyter, Berlin (2003)
20. Hajič, J., Hladka, B.V., Panevová, J., Hajičová, E., Sgall, P., Pajas, P.: Prague Dependency Treebank 1.0. LDC (2001) 2001T10
21. Kromann, M.T.: The Danish Dependency Treebank and the DTAG treebank tool. In: Second Workshop on Treebanks and Linguistic Theories, Växjö, Sweden, pp. 217–220 (2003)
22. Daum, M., Foth, K., Menzel, W.: Automatic transformation of phrase treebanks to dependency trees. In: Proceedings of the 4th International Conference on Language Resources and Evaluation, Lisbon, Portugal (2004)
23. Forst, M., Bertomeu, N., Crysmann, B., Fouvry, F., Hansen-Schirra, S., Kordoni, V.: Towards a dependency-based gold standard for German parsers: The TiGer Dependency Bank. In: Proceedings of LINC 2004, Geneva, Switzerland (2004)
24. Hudson, R.: Word Grammar. Basil Blackwell, Oxford (1984)
25. Engel, U.: Deutsche Grammatik. Groos, Heidelberg (1988)
26. Lobin, H.: Koordinationssyntax als prozedurales Phänomen. Studien zur deutschen Grammatik, vol. 46. Narr, Tübingen (1993)
27. Osenova, P., Simov, K.: BTB-TR05: BulTreebank Stylebook. Technical Report 05, BulTreeBank Project (2004)

A Learnable Representation for Syntax Using Residuated Lattices

Alexander Clark

Department of Computer Science,
Royal Holloway, University of London
alexc@cs.rhul.ac.uk

Abstract. We propose a representation for natural language syntax based on the theory of residuated lattices: in particular on the Galois lattice between contexts and substrings, which we call the syntactic concept lattice. The natural representation derived from this is a richly structured context sensitive formalism that can be learned using a generalisation of distributional learning. In this paper we define the basic algebraic properties of the syntactic concept lattice, together with a representation derived from this lattice and discuss the generative power of the formalism. We establish some basic results which show that these representations, because they are defined language theoretically, can be inferred from information about the set of grammatical strings of the language. We also discuss the relation to other grammatical formalisms notably categorial grammar and context free grammars. We claim that this lattice based formalism is plausibly both learnable from evidence about the grammatical strings of a language and may be powerful enough to represent natural languages, and thus presents a potential solution to the central problem of theoretical linguistics.

1 Introduction

Given arbitrary amounts of information about a language, how can we construct a representation for that language? In formal grammar, we define a class of representations for languages, and a function from these representations into the set of languages. Thus we have the class of Context Free Grammars (CFGs), and for each CFG, G, we have the language $L(G)$. Given a grammar, there is an efficient procedure for determining whether a string $w \in L(G)$. However, the inverse problem is hard: given information about the strings that are in some language L, there are no good procedures for finding a context free grammar G such that $L = L(G)$. Deciding what the non-terminals of the grammar should be is hard for linguists to do; it requires tenuous lines of argument from unreliable constituent structure tests, and disputes cannot be resolved easily or empirically. Moreover, the end results are inadequate: no one has ever produced a descriptively adequate generative grammar for any natural language. Even in English, the most well studied language, the "correct" analysis is often unclear. The problem is that the representation is radically under-determined by the evidence. Since there are many possible CF grammars for any CF language, indeed

P. de Groote, M. Egg, and L. Kallmeyer (Eds.): FG 2009, LNAI 5591, pp. 183–198, 2011.
© Springer-Verlag Berlin Heidelberg 2011

infinitely many for any infinite language, the task is ill defined without additional constraints, or an "evaluation procedure" in Chomskyan terms. The mechanisation of this process – i.e. the problem of grammatical inference – is thus also extremely hard. Even learning regular grammars (i.e. non-deterministic finite state automata) is hard computationally, even in quite a benign learning model [1]; and therefore learning CFGs, or Tree-adjoining grammars is also hard.

In this paper, we show that it is possible to define alternative representations that have very attractive properties from a learning point of view. The key insight is contained in [2]: if the representational primitives of the model are defined language theoretically, then the model will be easy to learn in an unsupervised way. Thus rather than defining a representation (like a CFG), and then defining a map from the representation to the language that it defines, we proceed in the opposite direction. We start by defining a map from the *language* to the *representation*. This reduces or eliminates the underdetermination of the representation by the language. Traditional algorithms for the inference of deterministic finite state automata exploit this approach implicitly through the Myhill-Nerode theorem; [2] exploit it explicitly through identifying non-terminals in a context free grammar with congruence classes of the language. However neither of these models are descriptively adequate for natural language syntax.

Here, we extend [3], and rather than basing our model on the congruence classes of the language, we base it on the *lattice* structure of those congruence classes: these form a residuated lattice that we call the *syntactic concept lattice*. This greatly enlarges the class of languages that can be represented. A remarkable consequence, which hints strongly that this approach is on the right track, is that the natural representation derived from these considerations is in fact not a context free formalism, but rather includes some non context free, mildly context sensitive languages. In addition it is capable of representing richly structured languages, which require structured non-terminals (i.e. augmented with feature structures) to be compactly represented by a context free grammar. It does not include all CFLs, but the examples of those that it does not are bizarre and do not correspond to phenomena in natural language.

The contributions of this paper are:

1. A definition of the syntactic concept lattice;
2. A proof that this is a residuated lattice;
3. The definition of a grammatical formalism based on this, which is very close to the learnable class of contextual binary feature grammars [3];
4. Some basic results that establish that these representations can be learned merely from information about which strings are in the language.

2   Contexts and Syntactic Concepts

Distributional learning [4] broadly conceived is the approach that tries to infer representations based on the "distribution" of strings. Given a finite non-empty alphabet Σ, we use Σ^* to refer to the set of all strings and λ to refer to the empty string. A context is just an ordered pair of strings that we write (l, r)

– l and r refer to left and right. We can combine a context (l, r) with a string u with a wrapping operation that we write $\odot$: so $(l, r) \odot u$ is defined to be lur. We will sometimes write f for a context (l, r). Given a formal language $L \subseteq \Sigma^*$ we can consider a relation between contexts $(l, r) \in \Sigma^* \times \Sigma^*$ and strings w given by $(l, r) \sim_L w$ iff $lwr \in L$. For a given string w we can define the *distribution* of that string to be the set of all contexts that it can appear in: $C_L(w) = \{(l, r) | lwr \in L\}$, equivalently $\{f | f \odot w \in L\}$. There is a special context (λ, λ): clearly $(\lambda, \lambda) \in C_L(w)$ iff $w \in L$. There is a natural equivalence relation on strings defined by equality of distribution: $u \equiv_L v$ iff $C_L(u) = C_L(v)$; this is called the syntactic congruence. We write $[u]$ for the congruence class of u. A learning algorithm based on this gave rise to the first linguistically interesting learnability result for context free languages [2]: this used the congruence classes to be the non terminals of a context free grammar together with the basic rule schemas $[uv] \rightarrow [u][v]$ and $[a] \rightarrow a$. In terms of the syntactic monoid, if X, Y, Z are elements of the syntactic monoid then $X \rightarrow YZ$ is a production iff $X = Y \circ Z$. Thus the algebraic properties of the monoid define the structure of the grammar directly.

We can also define the natural dual equivalence relation for contexts $(l, r) \equiv_L (l', r')$ iff for all w, $lwr \in L$ iff $l'wr' \in L$, and we write $[l, r]$ for the equivalence class of the context (l, r) under this relation.

3   Concept Lattice

It was realised first by [5] that the relation $\sim_L$ forms a Galois connection between sets of contexts and sets of strings. Galois lattices have been studied extensively in computer science and data mining under the name of Formal Concept Analysis [6] and these distributional lattices have been used occasionally in NLP for lexical analysis, e.g. [7,8]. For a modern treatment of Galois lattices and lattice theory in general see [9].

For a given language L we can define two polar maps from sets of strings to sets of contexts and vice versa. Given a set of strings S we can define a set of contexts S' to be the set of contexts that appear with every element of S.

$$S' = \{(l, r) : \forall w \in S \ lwr \in L\} \tag{1}$$

Dually we can define for a set of contexts C the set of strings C' that occur with all of the elements of C

$$C' = \{w : \forall (l, r) \in C \ lwr \in L\} \tag{2}$$

We define a syntactic concept to be an ordered pair of a set of strings S and a set of contexts C, written $\langle S, C \rangle$, such that $S' = C$ and $C' = S$. A set of strings (contexts) is closed iff $S = S''$ ($C = C''$). Note that for any sets $S''' = S'$ and $C''' = C'$. Thus for any set of strings S we can define a concept $\mathcal{C}(S) = \langle S'', S' \rangle$, and similarly for any set of contexts C, we can define $\mathcal{C}(C) = \langle C', C'' \rangle$.

We can define a partial order on these concepts where:

$$\langle S_1, C_1 \rangle \le \langle S_2, C_2 \rangle \text{ iff } S_1 \subseteq S_2.$$

$S_1 \subseteq S_2$ iff $C_1 \supseteq C_2$. We can see that $\mathcal{C}(L) = \mathcal{C}(\{(\lambda, \lambda)\})$, and clearly $w \in L$ iff $\mathcal{C}(\{w\}) \le \mathcal{C}(\{(\lambda, \lambda)\})$. We will drop some brackets to improve legibility.

Definition 1. *The concepts of a language L form a complete lattice $\mathfrak{B}(L)$, called the syntactic concept lattice, where $\top = \mathcal{C}(\Sigma^*)$, $\bot = \mathcal{C}(\Sigma^* \times \Sigma^*)$, where $\langle S_x, C_x \rangle \wedge \langle S_y, C_y \rangle$ is defined as $\langle S_x \cap S_y, (S_x \cap S_y)' \rangle$ and $\vee$ dually as $\langle (C_x \cap C_y)', C_x \cap C_y \rangle$.*

It is easy to verify that these operations satisfy the axioms of a lattice.

Figure 1 shows the syntactic concept lattice for the regular language $L = \{(ab)^*\}$. Note that though L is infinite, the lattice $\mathfrak{B}(L)$ is finite and has only 7 concepts.

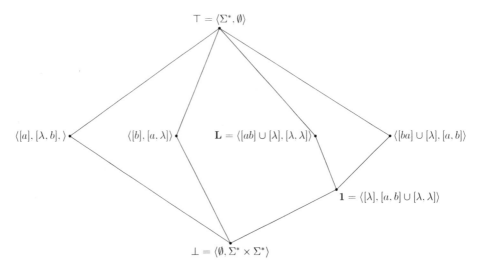

Fig. 1. The Hasse diagram for the syntactic concept lattice for the regular language $L = \{(ab)^*\}$. Each concept (node in the diagram) is an ordered pair of a set of strings, and a set of contexts. We write $[u]$ for the equivalence class of the string u, $[l, r]$ for the equivalence class of the context (l, r).

4   Monoid Structure

In addition to the lattice structure of $\mathfrak{B}(L)$, we can also give it a monoid structure. We define the concatenation of two concepts as follows:

Definition 2. $\langle S_x, C_x \rangle \circ \langle S_y, C_y \rangle = \langle (S_x S_y)'', (S_x S_y)' \rangle$

It is easy to verify that the result is a concept, that this operation is associative, and that $\mathbf{1} = \mathcal{C}(\lambda)$ is the unit and that this monoid operation respects the partial order of the lattice, in that if $X \leq Y$, then $W \circ X \circ Z \leq W \circ Y \circ Z$. This is therefore a lattice-ordered monoid. The left part of Table 1 shows this operation for the language $L = \{(ab)^*\}$. Moreover, we can define two residual operations as follows. We extend the operation $\odot$ to contexts as $(l, r) \odot (l', r') = (ll', r'r)$, so $(f_1 \odot f_2) \odot w = f_1 \odot (f_2 \odot w)$ for two contexts f_1, f_2 and a string w. We extend it to sets in the natural way. So for example, $C \odot (\lambda, S) = \{(l, yr) | (l, r) \in C, y \in S\}$.

Definition 3. *Suppose* $X = \langle S_x, C_x \rangle$ *and* $Y = \langle S_y, C_y \rangle$ *are concepts. Then define the residual* $X/Y = \mathcal{C}(C_x \odot (\lambda, S_y))$ *and* $Y \setminus X = \mathcal{C}(C_x \odot (S_y, \lambda))$

These are unique, and satisfy the following conditions:

Lemma 1. $Y \leq X \setminus Z$ *iff* $X \circ Y \leq Z$ *iff* $X \leq Z/Y$.

Proof. Suppose $X \leq Z/Y$; then $S'_x = C_x \supseteq C_z \odot (\lambda, S_y)$. Therefore $(S_x S_y)' \supseteq C_z$, and so $(S_x S_y)'' \subseteq C'_z = S_z$, and so $X \circ Y \leq Z$. Conversely suppose that $X \circ Y \leq Z$. Then we know that $lxyr \in L$ for any $x \in S_x, y \in S_Y, (l, r) \in C_z$. Therefore x must have all of the contexts of the form (l, yr), i.e. $C_x \supseteq C_z \odot (\lambda, S_y)$, and so $X \leq Z/Y$. Exactly similar arguments hold for $X \setminus Z$.

Therefore the syntactic concept lattice is a residuated lattice [10]. The map that takes $S \rightarrow \langle S'', S' \rangle$ for arbitrary sets of strings is a $\mathbf{1}, \vee, \circ$-homomorphism from the "free" residuated lattice of the powerset of Σ^*; but not a homomorphism of $\wedge$. Every language, computable or not, has a unique well defined syntactic concept lattice, which we can use as the basis for a representation that will be accurate for a certain class of languages.

Table 1. The concatenation operation. On the left is the operation for the language $L = \{(ab)^*\}$. $R = \mathcal{C}(ba), A = \mathcal{C}(A)$ etc. On the right is the concatenation operation for the partial lattice for the context free example described in Figure 2. We write A for $\mathcal{C}(a)$ and similarly for $AAB, AB, \ldots$. Note that $\mathbf{1}$ is an identity, but that this operation is not associative: $(A \circ A) \circ B \neq A \circ (A \circ B)$. If the result of an operation is $\top$ then the operation is vacuous.

∘	T	L	1	R	A	B	⊥
T	T	T	T	T	T	T	⊥
L	T	L	L	T	A	T	⊥
1	T	L	1	R	A	B	⊥
R	T	T	R	R	T	B	⊥
A	T	T	A	A	T	L	⊥
B	T	B	B	T	R	T	⊥
⊥	⊥	⊥	⊥	⊥	⊥	⊥	⊥

∘	T	AAB	A	ABB	B	AB	1	⊥
T	T	T	T	T	T	T	T	⊥
AAB	T	T	T	T	AB	T	AAB	⊥
A	T	T	T	AB	AB	AAB	A	⊥
ABB	T	T	T	T	T	T	ABB	⊥
B	T	T	T	T	T	T	B	⊥
AB	T	T	T	T	ABB	T	AB	⊥
1	T	AAB	A	ABB	B	AB	1	⊥
⊥	⊥	⊥	⊥	⊥	⊥	⊥	⊥	⊥

5   Partial Lattice

The lattice $\mathfrak{B}(L)$ will be finite iff L is regular. If we wish to find finite representations for non-regular languages we will wish to only model a fraction of this lattice. We can do this by taking a finite set of contexts $F \subset \Sigma^* \times \Sigma^*$ and constructing a lattice using only these contexts and all strings Σ^*. This give us a finite lattice $\mathfrak{B}(L, F)$, which will have at most $2^{|F|}$ elements. We can think of F as being a set of *features*, where a string w has the feature (context) (l, r) iff $lwr \in L$.

Definition 4. *For a language L and a set of context $F \subseteq \Sigma^* \times \Sigma^*$, the partial lattice $\mathfrak{B}(L, F)$ is the lattice of concepts $\langle S, C \rangle$ where $C \subseteq F$, and where $C = S' \cap F$, and $S = C'$.*

For example, consider everyone's favourite example of a context free language: $L = \{a^n b^n | n \geq 0\}$. The concept lattice for this language is infinite: among other concepts we will have an infinite number of concepts corresponding to a^i for each integral value of i. If we take the finite set of the six contexts $F = \{(\lambda, \lambda), (\lambda, b), (a, \lambda), (aab, \lambda), (\lambda, abb), (\lambda, ab)\}$, then we will end up with the finite lattice shown on the left hand side of Figure 2.

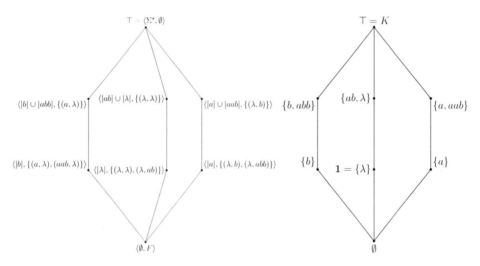

Fig. 2. The Hasse diagram for the partial lattice for $L = \{a^n b^n | n \geq 0\}$, with 6 contexts. The top element $\top$ contains all the strings, and the bottom element contains no strings. The empty string λ is below the language concept, as it has the context (λ, ab). On the left we have the true lattice; on the right we have the lattice as inferred from a small set of strings.

We can define a concatenation operation as before

$$\langle S_1, C_1 \rangle \circ \langle S_2, C_2 \rangle = \langle ((S_1 S_2)' \cap F)', (S_1 S_2)' \cap F \rangle$$

This is now however no longer a residuated lattice as the $\circ$ operation is no longer associative, there may not be an identity element, nor are the residuation operations well defined. The right hand side of Table 1 shows the concatenation operation for the lattice of Figure 2. We lose the nice properties of the residuated lattice, but the concatenation is still monotonic with respect to the lattice order – algebraically this partial lattice is a po-groupoid or lattice-ordered magma.

Given this partial lattice, we can define a representation that will map strings to concepts of this lattice. Clearly we have the map $w \mapsto \mathcal{C}(w)$ where each string is mapped to the correct concept. The concept $\mathcal{C}(w)$ is a finite representation of the distribution of the string w, of $C_L(w)$. Since there may be infinitely many congruence classes in the language, we cannot hope for this representation to be helpful for all strings, but only for some of the strings. For many strings, the representation we will get will be $\top$, which is vacuous – it tells us nothing about the string. However, what we are interested in is predicting one specific context, (λ, λ), and it may be that we can predict that context exactly.

The lattice $\mathfrak{B}(L, F)$ will consist of only finitely many concepts, so suppose we have a list of them and that we can compute the operations $\wedge$, $\vee$, and $\circ$. Then we can define a recursive function that uses the lattice to compute an estimate of $\mathcal{C}(w)$.

Definition 5. *For any language L and set of contexts F, which therefore defines a lattice $\mathfrak{B}(L, F)$, we define $\phi : \Sigma^* \to \mathfrak{B}(L, F)$ recursively by*

- *$\phi(\lambda) = \mathcal{C}(\lambda)$*
- *$\phi(a) = \mathcal{C}(a)$ for all $a \in \Sigma$, (i.e. for all $w, |w| = 1$)*
- *for all w with $|w| > 1$,*

$$\phi(w) = \bigwedge_{u,v \in \Sigma^+ : uv = w} \phi(u) \circ \phi(v) \tag{3}$$

or alternatively if we take $w = a_1 \ldots a_n$,

$$\phi(w) = \bigwedge_{i=1}^{n-1} \phi(a_1 \ldots a_i) \circ \phi(a_{i+1} \ldots a_n) \tag{4}$$

This is a recursive definition, that can be efficiently computed in $\mathcal{O}(|w|^3)$ time using dynamic programming techniques, analogously to the CKY algorithm. Equation 3 needs some explanation. In $\mathfrak{B}(L)$, $\mathcal{C}(u) \circ \mathcal{C}(v) = \mathcal{C}(uv)$. This is not always true in $\mathfrak{B}(L, F)$, but we can establish that $\mathcal{C}(u) \circ \mathcal{C}(v) \geq \mathcal{C}(uv)$. Thus if $w = uv$ we know that $\mathcal{C}(w)$ will be less than $\mathcal{C}(u) \circ \mathcal{C}(v)$, and that this will be true for all u, v such that $uv = w$. Since it is less than all of them, it will be less than the $\wedge$ over them all since meet is a greatest lower bound. So we know that $\mathcal{C}(w) \leq \bigwedge_{u,v} \mathcal{C}(u) \circ \mathcal{C}(v)$. So $\phi(w)$ simply recursively computes this upper bound on $\mathcal{C}(w)$. In the worst case this will just trivially give $\top$, but we hope that this upper bound will often be tight so that $\phi(w) = \mathcal{C}(w)$. This may appear to be a slightly circular definition, as it is indeed: however, as we shall see we can infer the algebraic structure of the lattice in a straightforward way from data, at which point it ceases to be circular.

Note that $\phi(w)$ aggregates information from every bracketing of w; since $\circ$ is not in general associative in the partial lattice, each bracketing may give a different result. Moreover, the formalism can combine information from different derivations which is more powerful: this is because $W \circ (X \wedge Y) \leq (W \circ X) \wedge (W \circ Y)$, but often without equality. This means that it can represent some context sensitive languages, and that the structural descriptions or derivations will no longer be trees, but rather directed acyclic graphs.

This map depends on L and F, but we can establish that it is always an upper bound:

Lemma 2. *For any language L and set of contexts F, for any string w, $\phi(w) \geq \mathcal{C}(w)$.*

This means that the set of contexts predicted will be a subset of the true set of features.

Proof. By recursion on length of w, using the fact that $\mathcal{C}(u) \circ \mathcal{C}(v) \geq \mathcal{C}(uv)$. Clearly it is true for $|w| = 1$ and $|w| = 0$, by construction $\mathcal{C}(w) = \phi(w)$. Suppose it is true for $|w| \leq k$. By definition $\phi(w)$ is a meet over the different elements $\phi(u) \circ \phi(v)$, where $w = uv$. By the inductive hypothesis, $\phi(u) \geq \mathcal{C}(u)$, and $\phi(v) \geq \mathcal{C}(v)$. Therefore, for each u, v we have that $\phi(u) \circ \phi(v) \geq \mathcal{C}(u) \circ \mathcal{C}(v) \geq \mathcal{C}(w)$. $\mathcal{C}(w)$ must be less than or equal to the greatest lower bound (meet) of all these $\phi(u) \circ \phi(v)$ which is $\phi(w)$.

Assuming that $(\lambda, \lambda) \in F$, we can define the language generated by this representation to be:

$$\hat{L} = L(\mathfrak{B}(L, F)) = \{w | \phi(w) \leq \mathcal{C}((\lambda, \lambda))\} \tag{5}$$

As a simple corollary we have:

Lemma 3. *For any language L and for any set of contexts F, $L(\mathfrak{B}(L, F)) \subseteq L$.*

We can define a class of languages $\mathcal{L}$ as the set of all languages L such that there is a finite set of contexts F such that $L = L(\mathfrak{B}(L, F))$. Since we have made no assumptions at all about L up to now, not even that is is computable, we will find that there are many languages where there is no finite set of contexts that define the language. $\mathcal{L}$ is clearly a countable class of recursive languages; we will discuss the language theoretic properties of this class in Section 7, but we will see that it does not correspond to the Chomsky hierarchy.

To recap: given the set of contexts F and the language L, we have a uniquely defined partial lattice $\mathfrak{B}(L, F)$, that as we shall see can be inferred from data about the language. This lattice can be considered as a representation that computes for every substring an estimate of the distribution of that substring. If this estimate predicts that the substring can occur in the context (λ, λ) then the string is in the language defined by this representation. ϕ is the recursive computation defined by the lattice that computes this estimate.

6   Learnability

These representations have been defined in language theoretic terms: that is to say, rather than focussing on the function from representation to language, we have focussed on defining a function from the language to the representation. It is therefore straightforward to find learning algorithms for this class. The lattice representation exactly mirrors a fragment of the syntactic concept lattice of the target language.

This paper is not focussed on the learning algorithms: see [3] for the first result along these lines. We will state some results that we hope make clear the fundamental tractability of the learning process. First, that if we increase the set of contexts then the language will increase. This, combined with Lemma 3, means that any sufficiently large context set will define the correct language.

Lemma 4. *If $F \subseteq G$ then $L(\mathfrak{B}(L, F)) \subseteq L(\mathfrak{B}(L, G))$*

We will prove this by establishing a more general lemma. We define the obvious map f from $\mathfrak{B}(L, G) \to \mathfrak{B}(L, F)$, as $\langle S, C \rangle \mapsto \langle (C \cap F)', C \cap F \rangle$. and the map back as $f^*(\langle S, C \rangle) = \langle S, S' \rangle$.[1]

Lemma 5. *For any language L, and two sets of contexts $F \subseteq G$, given the two computational maps $\phi_F : \Sigma^* \to \mathfrak{B}(L, F)$ and $\phi_G : \Sigma^* \to \mathfrak{B}(L, G)$, then for all w, $f(\phi_G(w)) \leq \phi_F(w)$.*

Intuitively this lemma says that the richer, bigger lattice will give us more specific, better predictions.

Proof. We will use two facts about the map f: $f(X \wedge Y) \leq f(X) \wedge f(Y)$, and $f(X \circ Y) \leq f(X) \circ f(Y)$. Again by recursion on the length of w; clearly if $|w| = 1$, $f(\phi_G(w)) = f(\mathcal{C}(w)_G) = \mathcal{C}(w)_F = \phi_F(w)$. We put a subscript to $\mathcal{C}()$ to indicate which lattice we are talking about. Inductive step: $f(\phi_G(w)) = f(\bigwedge_{u,v} \phi_G(u) \circ \phi_G(v)) \leq \bigwedge_{u,v} f(\phi_G(u) \circ \phi_G(v)) \leq \bigwedge_{u,v} f(\phi_G(u)) \circ f(\phi_G(v)) \leq \bigwedge_{u,v} \phi_F(u) \circ \phi_F(v) = \phi_F(w)$.

The second lemma is that we can infer $\mathfrak{B}(L, F)$ from a finite amount of data. Given a finite set of strings K we can define the lattice $\mathfrak{B}(K, L, F)$ in the obvious way:

Definition 6. *Given a set of strings K and a set of contexts F and a language L, define $\mathfrak{B}(K, L, F)$ to be the complete lattice formed by the ordered pairs $\langle S, C \rangle$, where $S \subseteq K$, $C \subseteq F$, and where $S' \cap F = C$, and $C' \cap K = S$.*

To avoid problems, let us be clear. The concepts are ordered pairs that consist of a finite set of strings, from K, and a finite set of contexts from F.

Definition 7. *Let $\langle S_1, C_1 \rangle$ and $\langle S_2, C_2 \rangle$ be in $\mathfrak{B}(K, L, F)$. So S_1, S_2 are subsets of K. Define a set of contexts $(S_1 S_2)' \cap F$. If this set of contexts is closed in $\mathfrak{B}(K, L, F)$, then we define the concatenation $\langle S_1, C_1 \rangle \circ \langle S_2, C_2 \rangle = \langle ((S_1 S_2)' \cap F)' \cap K, (S_1 S_2)' \cap F \rangle$. Otherwise it is undefined.*

---

[1] The two maps f, f^* form a residuated pair in that $f(X) \leq Y$ iff $X \leq f^*(Y)$, for any $X \in \mathfrak{B}(L, G)$ and $Y \in \mathfrak{B}(L, F)$.

Example 1. Suppose $L = \{(ab)^*\}$ and $F = \{(\lambda, \lambda), (a, \lambda), (\lambda, b), (a, b)\}$, and $K = \{\lambda, a, b\}$.. $\mathfrak{B}(K, L, F)$ will have 5 concepts $\top, \bot, \mathcal{C}(a), \mathcal{C}(b), \mathcal{C}(\lambda)$. If we try to compute $\mathcal{C}(a) \circ \mathcal{C}(b)$, everything is fine. We have ab which has contexts (λ, λ) which is closed, so $\mathcal{C}(a) \circ \mathcal{C}(b) = \mathcal{C}((\lambda, \lambda))$. But if we try to compute $\mathcal{C}(b) \circ \mathcal{C}(a)$, we will get the singleton set of strings $\{ba\}$, which has context set $\{(a, b)\}$ which while it is closed in $\mathfrak{B}(L, F)$ is not closed in $\mathfrak{B}(K, L, F)$.

Definition 8. *The lattice $\mathfrak{B}(K, L, F)$ is closed under concatenation if for every pair of concepts X, Y, the concatenation $X \circ Y$ is defined.*

Note that $\mathfrak{B}(L, F)$ is exactly the same as $\mathfrak{B}(\Sigma^*, L, F)$. So in order to compute $\mathfrak{B}(K, L, F)$ we need to know which strings in $F \odot KK$ are in L; computationally then the algorithm is given F, K and the finite set of strings $(F \odot KK) \cap L$.

Lemma 6. *For any L, F, there is a finite set K such that $\mathfrak{B}(K, L, F)$ is closed under concatenation and is isomorphic to the $\mathfrak{B}(L, F)$.*

Proof. (Sketch) For each concept in the finite set of concepts $\{\mathcal{C}(u) \in \mathfrak{B}(L, F) | u \in \Sigma^*\}$ pick one such string u. For each pair of concepts X, Y and context f where f is not in $X \circ Y$ pick a pair of strings u, v such that $u \in X, v \in Y$ and uv does not have the context f.

Moreover as we increase the set of strings K that we are basing the lattice on, the language defined by the lattice will decrease. Intuitively, as we increase the sample size the set of contexts shared by all samples in a given set will only decrease; thus the predicted set of features will decrease, and ϕ will move higher in the lattice, thus reducing the language.

Definition 9. *If $J \subset K$ we define the map g from $\mathfrak{B}(J, L, F)$ to $\mathfrak{B}(K, L, F)$, i.e. from the smaller lattice to the larger lattice as the map that takes $\langle S, C \rangle$ to $\langle C' \cap K, C \rangle$, and the map g^* from $\mathfrak{B}(K, L, F)$ to $\mathfrak{B}(J, L, F)$, by $\langle S, C \rangle \mapsto \langle S \cap J, (S \cap J)' \cap F \rangle$.*

This is well-defined since $S \cap J$ will be closed in $\mathfrak{B}(J, L, F)$.

Lemma 7. *For all J, K closed, $J \subset K$, and for all strings w; we have that $g(\phi_J(w)) \leq \phi_K(w)$.*

Proof. (Sketch) Again by induction on length of w. Both J and K include the basic elements of Σ and λ. Suppose true for all w of length at most k, and take some w of length $k + 1$. We use some inequalities for g that we do not prove here.

$$\phi_K(w) = \bigwedge_{u,v} \phi_K(u) \circ \phi_K(v)$$

$$\geq \bigwedge_{u,v} g(\phi_J(u)) \circ g(\phi_J(v))$$

$$\geq \bigwedge_{u,v} g(\phi_J(u) \circ \phi_J(v))$$

$$\geq g\left(\left(\bigwedge_{u,v} \phi_J(u)\right) \circ \phi_J(v)\right) = g(\phi_J(w))$$

As a corollary we therefore have:

Lemma 8. *For any L, F and any sets of strings J, K s.t $\Sigma \cup \{\lambda\} \subseteq J \subseteq K$ and where both $\mathfrak{B}(J, L, F)$ and $\mathfrak{B}(K, L, F)$ are closed under concatenation, $L(\mathfrak{B}(J, L, F)) \supseteq L(\mathfrak{B}(K, L, F)) \supseteq L(\mathfrak{B}(L, F))$.*

Finally we note that there are efficient scalable algorithms for computing these lattices and identifying the frequent concepts; see for example [11]. Thus, for any sufficiently large set of contexts the lattice $\mathfrak{B}(L, F)$ will define the right language; there will be a finite set of strings K such that $\mathfrak{B}(K, L, F)$ is isomorphic to $\mathfrak{B}(L, F)$, and there are algorithms to construct these lattices from K, F and information about $\sim_L$.

This is still some way from a formal polynomial learnability result. Since the class of languages is suprafinite, we cannot get a learnability result without using probabilistic assumptions, which takes us out of the scope of this paper, but see the related learnability result using a membership oracle in a non probabilistic paradigm in [3]. Note however that the monotonicity lemmas in the current approach (Lemmas 5 and 7) are exactly the opposite of the monotonicity lemmas in [3].

Space does not permit a full example, but consider the CF language $L = \{a^n b^n | n \geq 0\}$. If $F = \{(\lambda, \lambda), (\lambda, b), (a, \lambda), (\lambda, abb), (aab, \lambda)\}$, it is easy to verify that $L(\mathfrak{B}(L, F)) = L$. If $K = \{\lambda, a, b, ab, aab, abb\}$, then $\mathfrak{B}(K, L, F)$ is isomorphic to $\mathfrak{B}(L, F)$. This is shown on the right hand side of Figure 2.

Algorithms based on heuristic approximations to this approach are clearly quite feasible: consider features to be sets of contexts of the form $(\Sigma^* a, b\Sigma^*)$ where $a, b \in \Sigma$; take all frequent contexts; take all frequent substrings; approximate the relation $(l, r) \sim_L w$ probabilistically using a clustering algorithm.

7 Power of the Representation

We now look at the language theoretic power of this formalism, and its relationship to other existing formalisms. We will define $\mathcal{L}$ to be the class of all languages L such that $L(\mathfrak{B}(L, F)) = L$ for some finite set of contexts F. The following propositions hold:

- $\mathcal{L}$ contains the class of regular languages.
- $\mathcal{L}$ contains some languages that are not context free.
- There are some context free languages that are not in $\mathcal{L}$.

The CBFG formalism is clearly closely related. However CBFGs only use the partial order and not the full lattice structure. Moreover the absence of unary rules for computing $\wedge$ limits the generative power: CBFGs assume the lattice is distributive. Note also the relation to Range Concatenation Grammars [12] and Conjunctive Grammars [13].

7.1   Categorial Grammars

We have shown that this concept lattice is a residuated lattice; the theory of categorial grammars is based largely on the theory of residuation [14]. It is worth considering the relation of the residuation operations in the concept lattice to the theory of categorial grammars. Clearly implication $\rightarrow$ in categorial grammar corresponds to $\leq$ in this algebraic framework. Every sequent rule such as $X \rightarrow Y/(X\backslash Y)$ can be stated in $\mathfrak{B}(L)$ as $\mathcal{C}(X) \leq \mathcal{C}(Y)/(\mathcal{C}(X)\backslash\mathcal{C}(Y))$: an inequality which is true for all residuated lattices. In terms of the sets of axioms, the concept lattice satisfies the axioms of the associative Lambek calculus $\mathbf{L}$. However from a logical point of view the calculus that we use is more powerful. Just as the Ajdukiewicz-Bar-Hillel calculus which only uses the symbols $\backslash, /$ was extended to the associative Lambek calculus with the symbol $\circ$, here we need to add the additional symbols $\wedge, \vee$ from the lattice operations, together with additional inference rules. Indeed it is the following inference rule that takes us out of the context free languages, since they are not closed under intersection:

$$\frac{\Gamma \rightarrow Y \qquad \Gamma \rightarrow Z}{\Gamma \rightarrow Y \wedge Z}$$

Given that [15] showed that the equational theory of residuated lattices is decidable, this means that the calculus derived from this formalism is also decidable.

However the approach taken in this paper is profoundly different: in the categorial grammar style formalism, the underlying model is the residuated lattice of all subsets of Σ^*, which is a different lattice to $\mathfrak{B}(L)$. The language is then defined equationally through the type assignments to the letters of Σ. Here the model is the syntactic concept lattice, and the language is defined algebraically through a direct representation of the algebraic structure of part of the lattice. It is not obvious therefore that the two approaches are potentially equivalent: we can convert the partial lattice into a set of equations, and define the full lattice to be the free residuated lattice generated by Σ and these equations, but it may not be possible to "lexicalise" these equations.

7.2   Context Free Grammars

We can also consider the relation to context free grammars. Using standard notation, for a CFG with a non-terminal N we can define the yield of N as

$$Y(N) = \{w \in \Sigma^* | N \overset{*}{\Rightarrow} w\}$$

and the distribution as

$$C(N) = \{(l, r) \in \Sigma^* \times \Sigma^* | S \overset{*}{\Rightarrow} lNr\}$$

It is natural to think that $\langle Y(N), C(N) \rangle \in \mathfrak{B}(L(G))$. This is sometimes the case but need not be; indeed from a learnability point of view this is the major flaw of CFGs: there is no reason why there should be any simple correspondence between the concepts of the language and the structure of the grammar. There are CFLs for which it is impossible to find CFGs where all of the non-terminals are syntactic concepts, indeed there are some where some of the non-terminals must correspond to concepts with context sensitive sets of strings.

However we can represent all CFLs that have context free grammars that have a "Finite Context Property":

Definition 10. *For a context free grammar G, a non-terminal N has the FCP iff there is a finite set of contexts $F(N)$ such that $\{w | \forall (l, r) \in F(N) \, lwr \in L\}$ is equal to $Y(N)$.*

If every non-terminal has the FCP, then it can be shown that the partial lattice with the union of all the $F(N)$, will define the same language as L. This means that for every non-terminal in the grammar, we must be able to pick a finite set of contexts, that suffice to pick out the strings that can be derived from that non-terminal: $F(N)$ will normally be a subset of $C(N)$. A single context normally suffices for the simple examples in this paper, but not in natural languages where lexical ambiguity and coordination mean that one may need several contexts to pick out exactly the right set of strings. So for example the context "I was —— .", does not pick out an adjective phrase as "a student" can also appear in that context.

7.3   Context Sensitive

We can also define some non-context free languages. In particular, we define a language closely related to the *MIX language* (consisting of strings with an equal number of a's, b's and c's in any order) which is known to be non context-free.

Example 2. Suppose we have a context sensitive language Let $M = \{(a, b, c)^*\}$, we consider the language $L = L_{abc} \cup L_{ab} \cup L_{ac}$ where $L_{ab} = \{wd | w \in M, |w|_a = |w|_b\}$, $L_{ac} = \{we | w \in M, |w|_a = |w|_c\}$, $L_{abc} = \{wf | w \in M, |w|_a = |w|_b = |w|_c\}$.

It is easy to see that this language is non context free. We define the set of contexts:

$$F = \{(\lambda, \lambda), (\lambda, d), (\lambda, ad), (\lambda, bd), (\lambda, e), (\lambda, ae), (\lambda, ce), (\lambda, f), (ab, \lambda), (ac, \lambda)\}$$

The resulting lattice will consist of 22 concepts, including top and bottom; and it can be shown easily that the lattice defines the language L. The key point is the existence of concepts $\mathcal{C}((\lambda, d))$ and $\mathcal{C}((\lambda, e))$ together with the non-trivial equation $\mathcal{C}((\lambda, d)) \wedge \mathcal{C}((\lambda, e)) = \mathcal{C}((\lambda, f))$.

8   Conclusion

We have presented a model for syntax as an alternative to CFGs and categorial grammars. These models have deep roots in the study of post-Harris structuralist linguistics in Europe and Russia; see the survey in [16].

We base the representation on an algebraic structure defined by the language: the syntactic concept lattice. Algebraic properties of this lattice can be converted directly into the "grammar": this largely removes the arbitrariness of the representation, and leads to inference algorithms, as well as consequences for decidability, the existence of canonical forms etc. Given the language, and the set of features, the representation is determined. Selecting the set of features is easy as any sufficiently large set of features will define the same language. If the set of features is too small, then the representation will define a subset of the correct language.

We have defined a representation which is finite, but it is not very compact since the representation may be exponentially large in the number of contexts. In many cases, the number of concepts is polynomially bounded, but it is clearly desirable to use a more efficient representation. Moreover in natural language, we need efficient ways of representing combinations of number, case and gender features in those languages that have them: such features cause an exponential explosion in the number of atomic categories required [17]. We omit, because of space limitations, discussion of these more efficient representations, which rely on representing the lattice using pairs of minimal elements, which allow a more abstract and compact representation.

One objection to this approach is that it doesn't produce the "right" results: the structural analyses that we produce do not necessarily agree with the structures that linguists use to describe sentences, for example in treebanks. But of course linguists don't know what the right structures are: tree structures are not empirical data, but rather theoretical constructs. It is certainly important that the models support semantic interpretation, but as we know from categorial grammar this does not require traditional constituent structure. Moreover, the residuated lattice structure we use has been studied extensively in the model theory for various logics; we therefore conjecture that it is possible to build semantic analysis on this lattice in a very clean way. Ultimately, we don't yet know what the "right" representations are, but we do know they are learnable. It seems reasonable therefore to start by looking for learnable representations.

Just as with context free grammars, we can define a derivation that will "prove" that the string is grammatical. With lattice grammars, rather than being trees, these will be directed acyclic graphs or hypergraphs, where the nodes are associated with a particular span or segment of the input string, and each node is labelled with a concept. Looking at the context sensitive example discussed above, it is clear that the derivation cannot in general be a tree. We think that minimal derivations will be particularly interesting for semantic interpretation: a derivation is minimal if the concept associated with a span cannot be replaced by a more general concept, while still maintaining the validity of the associated DAG. In this case, classical cases of lexical and syntactic ambiguity will give rise to structurally distinct minimal derivations. In general, there will not be a one-to-one map between derivations and semantic interpretations: rather there will be what is called in CG parsing "spurious ambiguity". If the set of features is large, then the number of possible derivations will increase. For a regular language, if we have one context out of each context class, we will have structural completeness, and every tree can give rise to a derivation. The other important difference with a CFG is that the labels in the structural descriptions of sentences are not atomic symbols, as is this case with a non-terminal in a CFG, but rather are elements of a lattice.

Converting an existing CFG into a lattice grammar is not always possible: however for CFGs that are models for natural languages it should be straightforward. All that is needed is to find a finite set of contexts that picks out the yield of each non-terminal: given a large sample of positive example trees generated by the CFG in question will give a large sample of $C(N)$ and $Y(N)$, for any non-terminal N. However this is not the right approach: we are not interesting in representing CFGs, but in representing natural languages, and the limitations of CFG as representations of natural language are very well known.

The most radical property of lattice grammars is this: lattice grammars in this approach are not written by human linguists, rather they are determined by the data. The normal activity of linguistics is that some interesting data are discovered and linguists will manually construct some appropriate analysis together with a grammar for the fragment of the language considered. There are normally many possible grammars and analyses. With lattice grammars, the data decides: there is no role for the linguist in deciding what the appropriate structural description for a sentence is. The linguist can decide what the contexts are, and how many are used, and thus can control to some extent the set of possible structural descriptions, but given the set of contexts, the rules and the possible analyses for individual sentences are fixed given agreement about the data.

There is an interesting link to the classic presentation found in traditional *descriptive* grammars, such as [18]. In such grammars, the syntactic properties of a word are described through their occurrences in sample sentences: the lattice approach allows a fairly direct translation of this description into a generative grammar: Consider for example adjectives in English, which typically appear in three positions attributive, predicative, and postpositive, and which accept degree modifiers and adverb modifiers [18, p.528]. Each of these properties can be associated with a context, writing the gap as "—" "They are —", "I saw some — people" , "There is someone —", "They are very —", "He is the most — person I know", "He was remarkably —". A word like "happy" can appear in all of the contexts.

These lattice based models potentially satisfy the three crucial constraints for a model of language [19]; they may be sufficiently expressive to represent natural languages compactly, they can be learned efficiently, and they do not posit a rich domain specific and evolutionary implausible language faculty. We suggest they are therefore worthy of further study.

Acknowledgments

I am very grateful to Rémi Eyraud and Amaury Habrard.

References

1. Angluin, D., Kharitonov, M.: When won't membership queries help? J. Comput. Syst. Sci. 50, 336–355 (1995)
2. Clark, A., Eyraud, R.: Polynomial identification in the limit of substitutable context-free languages. Journal of Machine Learning Research 8, 1725–1745 (2007)
3. Clark, A., Eyraud, R., Habrard, A.: A polynomial algorithm for the inference of context free languages. In: Clark, A., Coste, F., Miclet, L. (eds.) ICGI 2008. LNCS (LNAI), vol. 5278, pp. 29–42. Springer, Heidelberg (2008)
4. Harris, Z.: Distributional structure. In: Fodor, J.A., Katz, J.J. (eds.) The Structure of Language, pp. 33–49. Prentice-Hall, Englewood Cliffs (1954)
5. Sestier, A.: Contribution à une théorie ensembliste des classifications linguistiques. In: Premier Congrès de l'Association Française de Calcul, Grenoble, pp. 293–305 (1960)
6. Ganter, B., Wille, R.: Formal Concept Analysis: Mathematical Foundations. Springer, Heidelberg (1997)
7. Popa-Burca, L.: On algebraic distributional analysis of romanian lexical units. In: Kay, M. (ed.) Abstracts of the 1976 International Conference on Computational Linguistics COLING, p. 54 (1979)
8. Basili, R., Pazienza, M., Vindigni, M.: Corpus-driven unsupervised learning of verb subcategorization frames. In: Lenzerini, M. (ed.) AI*IA 1997. LNCS, vol. 1321, pp. 159–170. Springer, Heidelberg (1997)
9. Davey, B.A., Priestley, H.A.: Introduction to Lattices and Order. Cambridge University Press, Cambridge (2002)
10. Jipsen, P., Tsinakis, C.: A survey of residuated lattices. Ordered Algebraic Structures, 19–56 (2002)
11. Choi, V.: Faster algorithms for constructing a galois/concept lattice. In: SIAM Conference on Discrete Mathematics 2006. University of Victoria, Canada (2006)
12. Boullier, P.: Chinese Numbers, MIX, Scrambling, and Range Concatenation Grammars. In: Proceedings of the 9th Conference of the European Chapter of the Association for Computational Linguistics (EACL 1999), pp. 8–12 (1999)
13. Okhotin, A.: Conjunctive grammars. Journal of Automata, Languages and Combinatorics 6(4), 519–535 (2001)
14. Moortgat, M.: Multimodal linguistic inference. Journal of Logic, Language and Information 5(3), 349–385 (1996)
15. Ono, H., Komori, Y.: Logics Without the Contraction Rule. The Journal of Symbolic Logic 50(1), 169–201 (1985)
16. van Helden, W.: Case and gender: Concept formation between morphology and syntax (II volumes). Studies in Slavic and General Linguistics, Rodopi, Amsterdam-Atlanta (1993)
17. Gazdar, G., Klein, E., Pullum, G., Sag, I.: Generalised Phrase Structure Grammar. Basil Blackwell, London (1985)
18. Huddleston, R., Pullum, G., Bauer, L.: The Cambridge grammar of the English language. Cambridge University Press, New York (2002)
19. Jackendoff, R.: Alternative minimalist visions of language. In: Proceedings of Chicago Linguistics Society, vol. (41) (2008)

Prior Knowledge in Learning Finite Parameter Spaces

Dorota Głowacka, Louis Dorard, Alan Medlar, and John Shawe-Taylor

Department of Computer Science,
University College London

Abstract. We propose a new framework for computational analysis of language acquisition in a finite parameter space, for instance, in the "principles and parameters" approach to language. The *prior knowledge multi-armed bandit* algorithm abstracts the idea of a casino of slot machines in which a player has to play machines in order to find out how good they are, but where he has some prior knowledge that some machines are likely to have similar rates of reward. Each grammar is represented as an arm of a bandit machine with the mean-reward function drawn from a Gaussian Process specified by a covariance function between grammars. We test our algorithm on a ten-parameter space and show that the number of iterations required to identify the target grammar is much smaller than the number of all the possible grammars that the learner would have to explore if he was searching exhaustively the entire parameter space.

1 Introduction

A major aspect of linguistic theory is to provide an explanation as to how children, after being exposed to limited data, acquire the language of their environment. The rise of "principles and parameters" [4] provided a new context for the study of language acquisition. In this approach, a class of languages can be viewed as fixed by parametric variation of a finite number of variables. Thus, acquisition of language (grammar)[1] amounts to fixing correctly the parameters of the grammar that the child is trying to learn. The notion of finite parametrisation of grammar can be applied to syntax [4], phonological stress systems [6] or even lexical knowledge [9]. In this paper, we will concentrate on the analysis of the "principles and parameters" framework as applied to stress systems [6]. This choice has been prompted mainly by two considerations. First, stress systems can be studied in relative independence of other aspects of grammars, i.e. syntax or semantics. Second, the parameters of metrical theory exhibit intricate interactions that exceed in complexity the syntactic parameters.

Starting with Gold's seminal paper [8], most research on learnability concentrates on the issue of convergence in the limit. The learner receives a sequence of

---

[1] In the remainder of this paper we will use the terms *language* and *grammar* interchangeably.

P. de Groote, M. Egg, and L. Kallmeyer (Eds.): FG 2009, LNAI 5591, pp. 199–213, 2011.
© Springer-Verlag Berlin Heidelberg 2011

positive examples from the target language. After each example the learner either stays in the same state (does not change any of the parameters) or moves to a new state (changes its parameter setting). If, after a finite number of examples the learner converges to the target language and never changes his guess, then the target language has been identified in the limit. In the Triggering Learning Algorithm (TLA) [7] two additional constraints were added: the single-value constraint, i.e. the learner can change only one parameter value at a time, and the greediness constraint, i.e. if, after receiving an example he cannot recognise, the learner changes one parameter and now can accept the new data, the learner retains the new parameter setting. The TLA is an online learning algorithm that performs local hill climbing. This algorithm, however, is problematic as positive-only examples can lead to local maxima, i.e. an incorrect hypothesis from which the learner cannot move, thus rendering the parameter space under consideration unlearnable. In order to acquire the target grammar, the learner has to start from very specific points in the parameter space.

[12], [13] model the TLA as a Markov chain and modify the TLA by replacing the local single-step hill climbing procedure with a simple Random Walk Algorithm (RWA). RWA renders the learning process faster and always converges to the correct target language irrespective of the initialisation of the algorithm. [12], [13] tested the system tested on a very small three-parameter system that produced 8 grammars, where each grammar consisted of only up to 18 acceptable examples. Following [12], [13]'s observation that a RWA greatly improves the accuracy and speed of the learning process, we propose a new way for computational analysis of language acquisition within the "principles and parameters" setting. Our algorithm is set within the general framework of multi-armed bandit problems. The *prior knowledge multi-armed bandit* problem abstracts the idea of a casino of slot machines in which a player has to play machines in order to find out how good they are, but where he has some prior knowledge that some machines are likely to have similar rates of reward. Each grammar is represented as "an arm of a bandit machine" with the mean-reward function drawn from a Gaussian Process specified by a covariance function between grammars. We test our algorithm on the metrical stress ten-parameter space [6], which gives rise to 216 possible stress systems (as not all parameters are independent). The use of the algorithm, however, can be easily extended to much larger systems. We also compare the performance of our algorithm to that of TLA and RWA and show that it "learns" the correct grammar faster than both TLA and RWA.

As emphasised by [7] and [12], [13], arriving at the correct parameter setting is only one aspect of the language acquisition problem. As noted by [3], an equally important point is how the space of possible grammars is "scattered" with respect to the primary language data. It is possible for two grammars to be so close to each other that it is almost impossible to separate them by psychologically realistic input data. This leads to the question of sample complexity [12], i.e. how many examples it will take to identify the target grammar. It is of not much use to the learner to be able to arrive at the correct target grammar within the limit if the time required to do so is exponentially long, which renders the learning process psychologically implausible. Thus, rather than concern ourselves with identifying the correct grammar, we will measure the number of errors made by the learner in acquiring the correct grammar. We will give experimental evidence that the number of iterations required to reach a state where the learner makes virtually no mistakes is smaller than the number of grammars to be explored. We will also consider the impact of variations in data distribution and the presence of noise in the data on the performance of the algorithm.

The remainder of the paper is organised as follows: first we briefly describe the basic tenets of metrical stress theory followed by a short description of the multi-armed bandit problem, where we introduce in more detail our arm selection procedure. Next, we describe how the procedure can be applied to learning parametrised grammars. Finally, we present experimental results on learning the ten-parameter space addressing a set of specific questions.

2   Metrical Stress Parameters and the Learning Problem

In this section, we describe the syllable structure and its relevance to stress assignment. Further, we present the stress parameters that we make a reference to throughout the rest of this paper. Lastly, we discuss how the data is presented to and assessed by the learner in our system.

2.1   Syllable Structure and Stress Assignment

We assume that the input to the learning process are words. One of the principles shared by most theories of stress systems is that stress is sensitive to representations built on projections from syllable structure. In many languages, stress is sensitive to syllable weight, or quantity. Thus, we also assume the prior operation of rules that convert the speech signal into words and smaller word segments, such as syllables.

In general, syllables can be divided into two parts: an onset (O) and a rhyme (R). The onset consists of the consonant(s) before the syllable peak, which is usually a vowel. The rhyme consists of the vowel and the consonant(s) following it. The rhyme can be further divided into two parts: the nucleus (N), i.e. the vowel, and the coda (C), i.e. the consonant(s) following it. It is generally agreed that the onset plays no part in stress assignment. However, in quantity-sensitive languages, the structure of the rhyme plays an important role in stress assignment (see [5] for possible counter examples). Syllables that have only one element in the nucleus position and no coda are classified as light (Fig. 1a) and as such do not attract stress. Syllables with two elements in the nucleus position count as heavy (Fig. 1c, d) and attract stress, while syllables with one element in the nucleus and at least one element in the coda position can count as either light or heavy (depending on the setting of parameter 6 below) (Fig. 1b).

Furthermore, we also assume that various acoustic cues that indicate phonological stress are mapped into one of three degrees of stress. The three levels of stress are primary stress (marked as 2), secondary stress (marked as 1), and lack

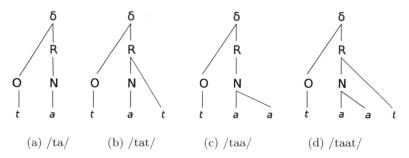

(a) /ta/   (b) /tat/   (c) /taa/   (d) /taat/

Fig. 1. Four examples of syllable with different rhyme structure (after [10]). δ signifies a syllable node; O, R, N and C represent the constituents of the syllable to which the segmental material is attached.

of stress (marked as 0). For the purpose of our analysis, we assume that every word must have a primary stress.

2.2   The Stress System Parameters

In metrical theory, stress patterns, and the corresponding differences between languages, are due to metrical structures built on the rhyme. The various possibilities of metrical structure construction can be expressed in terms of a series of binary parameters. In our analysis, we consider a 10-parameter model with the following parameters [6]:

- P1: The word-tree is strong on the left/right;
- P2: Feet are binary/unbounded;
- P3: Feet are built from left/right;
- P4: Feet are strong on the left/right;
- P5: Feet are quantity sensitive/insensitive;
- P6: Feet are quantity sensitive to the rhyme/nucleus;
- P7: A strong branch of the foot must/must not itself branch;
- P8: There is/is not an extrametrical syllable;
- P9: It is extrametrical on the left/right;
- P10: Feet are/are not non-iterative.

If all the parameters were independent, then we would have $2^{10} = 1024$ possible grammars. However, due to built-in dependencies, there only 216 distinct stress systems (see [6] for more details).

Let us consider the effect that different parameter settings can have on language structure. For example, P1 tells us where in the word the main stress should fall. If P1 is set to left, then the main stress will fall on the initial syllable, as in Latvian or Hungarian, if, however, we set P1 to right, then the main stress will fall on the final syllable, as in French or Farsi. In many languages, secondary stress can also be observed. In such languages, syllables are first grouped together into feet and every foot receives a stress. If a language has feet, a number of other parameters come into play. P2 allows feet to be at most binary or else unbounded. Selecting binary feet will give an alternating pattern of weak (with stress level 0) and strong (stress level 1 or 2) syllables. We must also set P3, which will trigger the direction of construction from left to right or from right to left. Further, we must also set P4, which allows each foot to be left dominated or right dominated. For example, Maranungku, spoken in Australia, [10] has the following setting P1[left], P2[binary], P3[left], P4[left], which gives rise to the following alternating pattern of stresses: 201, 2010, etc. On the other hand, Warao, spoken in Venezuela, [10] has the following setting: P1[right], P2[binary], P3[right], P4[left], which results in the following stress pattern: 020, 01020, 10101020, 010101020.

2.3   Inclusion of Prior Knowledge

In [7] and [12], [13], the transition probabilities from one parameter setting state to another are calculated by counting the number of overlapping input data between each grammar corresponding to each parameter setting. We consider this to be an unrealistic model in that it is not clear how the learner would be able to assess this overlap without knowledge of the grammars and the sentence frequencies. We prefer to work with a weaker assumption of the prior knowledge that learners are equipped with, namely that learners are able to assess similarity of grammars by the *Hamming distance* between their parameter vectors. This accords with the expectation that the entries in the parameter vector control aspects of the production of sentences that involve varying levels of processing by the learner. Hence, our conjecture is that the parameter settings described above have cognitive correlates that enable the learner to compute the *Hamming distance* between the grammars. One of the questions addressed by the experiments in this paper (and answered in the affirmative) is whether this prior knowledge will be sufficient to enable subjects to learn to identify the correct grammar.

2.4   The Learning Algorithm

Our learning procedure is partly inspired by the TLA [7]. However, contrary to [7], we do not obey the single-value constraint or the greediness constraint. Our learning algorithm can be summarised as follows:

- Step 1 [Initialise]: Start at some random point in the finite space of possible parameter settings and specify a grammar hypothesis.
- Step 2 [Process input data]: Receive n number of positive example words from the target grammar (g_t). The words are drawn at random from a fixed probability distribution.
- Step 3 [Learnability on error detection]: Check if the currently hypothesised grammar (g_h) can generate the input data and receive a reward r ranging from 0 to 1. $r = 0$ corresponds to a situation, where none of the n words can be found in the hypothesis grammar, while if $r = 1$, all the n words are

analysable in the currently hypothesised grammar. The reward function is calculated as follows:

$$r = \frac{\sum_{w_i \in g_h}^{n} pr_h(w_i)}{\sum_{w_i \in g_h}^{n} pr_h(w_i) + \sum_{w_i \notin g_h \cap w_i \in g_t}^{n} pr_t(w_i)} \tag{1}$$

where $pr(w_i)$ is the probability of the ith word.
- Step 4 [Update] Update distributions over possible grammars based on reward.
- Step 5 [Grammar selection]: Stay in the current hypothesis state or 'jump' to a new parameter setting (the new grammar selection procedure is described in detail in Sec. 4). The newly hypothesised grammar does not necessarily have to allow the learner to analyse all or any of the n input examples.

The learning process is completed when after some iteration m, the learner ceases to make any errors, i.e. at virtually every iteration after iteration m the reward at step 3 is always $r = 1$, and the grammar selected at step 5 is always the target grammar.

3   The Multi-armed Bandit Problem

The multi-armed bandit problem is an analogy with a traditional slot machine, known as a one-armed bandit, but with multiple arms. In the classical bandit scenario, the player, after playing an arm selected from a finite number of arms, receives a reward. The player has no initial knowledge about the arms, and attempts to maximise the cumulative reward through repeated plays. It is assumed that the reward obtained when playing an arm i is a sample from an unknown distribution R_i with mean μ_i. The optimal playing strategy S^*, i.e. the strategy that yields maximum cumulative reward, consists in always playing an arm i^* such that $i^* = \text{argmax}_i \mu_i$. The expected cumulative reward of S^* at time t would then be $t\mu_{i^*}$. The performance of a strategy S is assessed by the analysis of its expected regret at time t, defined as the difference between the expected cumulative reward of S^* and S at time t.

A good strategy requires to optimally balance the learning of the distributions R_i and the exploitation of arms which have been learnt as having high expected rewards. Even if the number of arms is finite and smaller than the number of experiments allowed so that it is possible to explore all the arms a certain number of times, this only gives probabilistic information about the best performing arms. The multi-armed bandit problem is concerned with the design and analysis of algorithms that can trade exploration and exploitation to achieve only a small regret for a finite set of independent arms. In our prior knowledge multi-armed bandit problem, we are interested in learning with many fewer trials through exploiting knowledge about similarities between different arms or, in our case, grammars. We will encode this information in a covariance function, hence assuming a prior Gaussian Process over reward functions. We now describe our learning algorithm.

3.1   The Prior Knowledge Multi-armed Bandit Algorithm

We consider space $\mathcal{X}$, whose elements will be referred to as arms. κ denotes a kernel between elements of $\mathcal{X}$. The reward after playing arm $\mathbf{x} \in \mathcal{X}$ is given by $f(\mathbf{x}) + \epsilon$, where $\epsilon \sim \mathcal{N}(0, \sigma_{noise}^2)$ and $f \sim \mathcal{GP}(\mathbf{0}, \kappa(\mathbf{x}, \mathbf{x}'))$ is chosen once and for all but is unknown. Arms played up to time t are $\mathbf{x}_1, \ldots, \mathbf{x}_t$ with rewards $y_1, \ldots, y_t$. The GP posterior at time t after seeing data $(\mathbf{x}_1, y_1), \ldots, (\mathbf{x}_t, y_t)$ has mean $\mu_t(\mathbf{x})$ with variance $\sigma_t^2(\mathbf{x})$.

Matrix C_t and vector $\mathbf{k}_t(\mathbf{x})$ are defined as follows:

$$(C_t)_{i,j} = \kappa(\mathbf{x}_i, \mathbf{x}_j) + \sigma_{noise}^2 \delta_{i,j} \tag{2}$$

$$(\mathbf{k}_t(\mathbf{x}))_i = \kappa(\mathbf{x}, \mathbf{x}_i) \tag{3}$$

$\mu_t(\mathbf{x})$ and $\sigma_t^2(\mathbf{x})$ are then given by the following equations (see [14]):

$$\mu_t(\mathbf{x}) = \mathbf{k}_t(\mathbf{x})^T C_t^{-1} \mathbf{y}_t \tag{4}$$

$$\sigma_t^2(\mathbf{x}) = \kappa(\mathbf{x}, \mathbf{x}) - \mathbf{k}_t(\mathbf{x})^T C_t^{-1} \mathbf{k}_t(\mathbf{x}) \tag{5}$$

As noted by [15], if no assumption is made on the nature of the reward function it may be arbitrarily hard to find an optimal arm. In their work, they assume that the mean-reward μ_k of a newly played arm k is a sample of a fixed distribution, and they characterise the probability of playing near-optimal arms. Others such as [11] and [2] assume that there exists a mean-reward function f and make further assumptions on the regularity of f. In the work of [11], arms are indexed in a metric space and the mean-reward is a Lipschitz function in this space. In the model of [2], arms lie in a generic topological space and f has a finite number of global maxima around which the function is locally Hoelder.

We assume in our model that the reward of an arm $\mathbf{x}$ is determined by a function f applied at point $\mathbf{x}$ to which Gaussian noise is added. The variance of the noise corresponds to the variability of the reward when always playing the same arm. In order to cope with large numbers of arms, our assumption will be that the rewards of arms are correlated. Thus, playing an arm 'close' to $\mathbf{x}$ gives information on the expected gain of playing $\mathbf{x}$. This can be modelled with a Gaussian Process: by default, we take the mean of the Gaussian Process prior to be $\mathbf{0}$, and we can incorporate prior knowledge on how correlated the arms are in the covariance function between arms. The covariance function can be seen as a kernel function, and specifies how 'close' or 'similar' two given arms are. Hence, we assume in our model that f is a function drawn from a Gaussian Process (GP).

If arms are indexed in $\mathbb{R}^d$, for example, the covariance function can be chosen to be a Gaussian kernel $\kappa(\mathbf{x}, \mathbf{x}') = \exp\left(-\frac{\|\mathbf{x} - \mathbf{x}'\|^2}{2\sigma^2}\right)$, whose smoothness σ is adjusted to fit the characteristic length scale which is assumed for f. In our framework, we are not limited to problems where arms are indexed in $\mathbb{R}^d$, but we can also consider arms indexed by any type of structured data as long as a kernel can be defined between the data points. For instance, in the parametric grammar learning problem, each grammar can be associated with an arm, so that looking for the optimal arm corresponds to looking for the optimal grammar given a certain criterion which is incorporated into the reward function.

Arm selection. The algorithm plays a sequence of arms and aims at optimally balancing exploration and exploitation. For this, arms are selected iteratively according to a UCB-inspired formula (see [1] for more details on UCB):

$$\mathbf{x}_{t+1} = \operatorname{argmax}_{\mathbf{x} \in \mathcal{X}} \{ f_t(\mathbf{x}) = \mu_t(\mathbf{x}) + B(t)\sigma_t(\mathbf{x}) \} \tag{6}$$

This can be interpreted as active learning where we want to learn accurately in regions where the function looks good, while ignoring inaccurate predictions elsewhere. The $B(t)$ term balances exploration and exploitation: the bigger it gets, the more it favours points with high $\sigma_t(\mathbf{x})$ (exploration), while if $B(t) = 0$, the algorithm is greedy. In the original UCB formula, $B(t) \sim \sqrt{\log t}$.

Although this arm selection method seems quite natural, in some cases, finding the maximum of the "upper confidence function" f_t may prove costly, particularly as we would expect the function to become flatter as iterations proceed, as the algorithm aims to explore regions only as our uncertainty about their reward offsets the difference in our estimated reward. For this reason we will consider an alternative arm selection method based on the sampling of functions from the GP posterior.

3.2   Application to the Grammar Learning Problem

As suggested above, the problem of grammar learning can be considered as a many-armed bandit problem and the Gaussian Process approach can be used with a covariance function/kernel which applies to different grammars. Learning consists in looking for the 'best' grammar, i.e. the one that maximises the reward function.

Let us denote by $\mathcal{X}$ the set of parametrised grammars. In the 'principles and parameters' framework, a grammar $\mathbf{x}$ is a binary vector of length d, where d is the number of parameters under consideration. In our case, $d = 10$. We need to define a kernel $\kappa(\mathbf{x}, \mathbf{x}')$/covariance function between grammars. In our experiments, we consider a Gaussian kernel that takes the form:

$$\kappa(\mathbf{x}, \mathbf{x}') = \exp\left(-\frac{\| \mathbf{x} - \mathbf{x}' \|^2}{2\sigma^2} \right) \tag{7}$$

where $\| \mathbf{x} - \mathbf{x}' \|^2$ is the *Hamming distance* between two grammars[2].

4   The Arm Selection Problem

The objective of the bandit algorithm is to minimise the regret over time, or in other words, to find as quickly as possible a good approximation of the maximum of f, $f(\mathbf{x}^*)$.

---

[2] Note that in this formulation, all parameters have equal effect on the produced data. As [6, p. 155, ft. 11] point out, theoretically, it is possible for a small change in the parameter setting to have large effects on the produced data and small changes to have small effects. This problem is not studied in [6]. However, if the parameters have a nonuniform effect on the output data, we can incorporate this information in the covariance function.

Let us define $f_t(\mathbf{x})$ by:

$$f_t(\mathbf{x}) = \mu_t(\mathbf{x}) + B(t)\sigma_t(\mathbf{x})$$

$$= \mathbf{k}_t(\mathbf{x})^T C_t^{-1}\mathbf{y}_t + B(t)\sqrt{\kappa(\mathbf{x},\mathbf{x}) - \mathbf{k}_t(\mathbf{x})^T C_t^{-1}\mathbf{k}_t(\mathbf{x})} \qquad (8)$$

Our approximation of $f(\mathbf{x}^*)$ at time t is $f(\mathbf{x}_{t+1})$ where $\mathbf{x}_{t+1} = \text{argmax}_{\mathbf{x}\in\mathcal{X}}\{f_t(\mathbf{x})\}$.

Here we replace the problem of finding the maximum of the function f by iterations of a simpler problem, which is to maximise the function f_t whose form is known. At each iteration we learn new information which enables us to improve our approximation of $f(\mathbf{x}^*)$ over time. In the case where $\kappa(\mathbf{x},\mathbf{x}) = 1$ for all $\mathbf{x}$ (e.g. Gaussian kernel and cosine kernel), $f_t(\mathbf{x})$ can be written as a function of $\mathbf{k}_t(\mathbf{x}) = \mathbf{k}$:

$$f_t(\mathbf{x}) = g(\mathbf{k}) = \mathbf{k}^T C_t^{-1}\mathbf{y}_t + B(t)\sqrt{1 - \mathbf{k}^T C_t^{-1}\mathbf{k}} \qquad (9)$$

4.1 Sampling Arm Selection

As we noted earlier, we would expect the upper confidence function to become flatter as iterations proceed, which would make it difficult to find its maximum. For this reason, we propose an alternative method for selecting the next arm rather than choosing the point that maximises the upper confidence function. Our strategy for selecting arms is to sample a function g from the posterior distribution and then select $\mathbf{x}_{t+1} = \text{argmax}_{\mathbf{x}\in\mathcal{X}}g(\mathbf{x})$. This implements sampling an arm with the probability that it is the maximum in the posterior distribution, and so implements a Bayesian approach to trading exploration and exploitation. We can interpolate between these methods by sampling a variable number K of functions $\mathbf{g}_1,\ldots,\mathbf{g}_K$ from the posterior and selecting $\mathbf{x}_{t+1} = \text{argmax}_{\substack{\mathbf{x}\in\mathcal{X}\\1\leq k\leq K}}\{g_k(\mathbf{x})\}$.

A naive sampling from the posterior distribution would require inverting a matrix indexed by the full grid. We avoid this by iteratively unveiling the posterior sample $g(\mathbf{x})$ only sampling points that are likely to lead to a maximal value. Since the function $g(\mathbf{x})$ is not expected to be flat, only a small number of samples should be required in practice. We would envisage selecting these samples by simple hill climbing heuristics that could work efficiently on the non-flat g, but in our experiments, we simply consider the enumeration of all 216 grammars.

Arm selection method:

1. Initialisation (t=1): $\mathbf{x}_1$ chosen randomly in $\mathcal{X}$. y_1 is the reward obtained when "playing" $\mathbf{x}_1$. The GP posterior after seeing the data $(\mathbf{x}_1, y_1)$ is sampled (f_1) to give iteration at time $t+1$ and $\mathbf{x}_2 = \text{argmax}\{f_t(\mathbf{x})\}$.
2. Iteration at time $t+1$: we have played $\mathbf{x}_1,\ldots,\mathbf{x}_t$ and have obtained rewards $y_1,\ldots,y_t$. The GP posterior is sampled to give f_t and $\mathbf{x}_{t+1} = \text{argmax}\{f_1(\mathbf{x})\}$.

We now describe the sampling method in detail, which returns the next selected arm $\mathbf{x}_{t+1}$. After seeing only the data $((\mathbf{x}_1, y_1)),\ldots,(\mathbf{x}_t, y_t)$, the posterior has mean $\mu_t^{(1)}$ and variance $\sigma_t^{(1)^2}$.

Algorithm 1. Sampling method

1: set $V = \{\}$ and $k = 1$
2: **for** $\mathbf{g}_i \in G$ **do**
3:    **if** $\mu_t^{(k)}(\mathbf{g}_i) + B(t)\sigma_t^{(k)}(\mathbf{g}_i) \geq \max(\{y_t\} \cup V)$ **then**
4:       sample from $\mathcal{N}(\mu_t^{(k)}(\mathbf{g}_i), \sigma_t^{(k)^2}(\mathbf{g}_i))$ and get v_k
5:       set $\mathbf{s}_k = \mathbf{g}_i$, $V = V \cup \{v_k\}$ and $k = k+1$
6:       the GP posterior after seeing the data $(\mathbf{x}_1, y_1), \ldots, (\mathbf{x}_t, y_t)$ with observation
         noise σ_{noise}^2 and $(\mathbf{s}_1, v_1), \ldots, (\mathbf{s}_k, v_k)$ without noise has mean $\mu_t^{(k+1)}$ and vari-
         ance $\sigma_t^{(k+1)}$
7:    **end if**
8: **end for**
9: return $\mathbf{s}_j$, where $v_j = \max(v_1, \ldots, v_{k-1})$

---

5   Experiments

The aim of the experiments is to investigate the following issues:

1. The learning process is completed in fewer iterations than the number of grammars under consideration.
2. The learning process is successful irrespective of the initial grammar selected in Step 1 of the learning algorithm.
3. The learning takes place in an on-line fashion, i.e. the number of errors made as the learning process progresses is gradually being reduced until it reaches 0.
4. The algorithm is robust with respect to the presence of noise and the input data distribution.

5.1   The Data

In our experiments, every input word consists of two parts: syllable representation followed by its stress assignment. In our analysis, we represent light syllables as L, syllables with a branching nucleus as N, and syllables with a branching rhyme as R. For example, a string of the form RLRL2010 represents a four-syllable word with primary stress on the initial syllable and secondary stress on the penultimate syllable; a pattern that can be found in, e.g. Icelandic or Czech. We consider words of up to a length of 7 syllables. This results from 2901 to 3279 words for each of the 216 possible grammars. Needless to say, a given word can belong to more than one grammar. The number of overlapping words between grammars ranges from 3 to 3189. The average number of overlapping words is 262.

As mentioned earlier, in our stress systems analysis we follow the principles of metrical stress theory [10] (further developed by [6]). Hayes [10] set the standard for much subsequent research in this field by bringing together data from around 400 natural languages and dialects and incorporating them into a unified metrical framework. The 216 grammars discussed here represent stress patterns of a wide range of natural languages from ancient languages (e.g. Latin) through well-known Indo-European languages (e.g. French or Czech) to native Australian or native America languages (e.g. Maranungku or Warao). A more exhaustive list of natural languages corresponding to each of the 216 grammars can be found in [10] and [6].

5.2   The Experiment Design and General Results

All the experiments reported below are averaged over 600 runs with a random starting point. This random initialisation allowed us to study the influence of the initialisation step of the learning process. Each run consisted of 400 iterations of the algorithm which we described in detail in Sec. 4.1. At each iteration the learner is presented with 5 words selected from the target grammar. The frequency with which each word is presented to the learner corresponds to the probability distribution of this word. Note that the learner is not allowed to use the particular words to inform his learning but only the average error of the currently hypothesised grammar on these words.

Below, we report results for the target grammar with the following parameter setting: P1[right], P2[binary], P3[right], P4[left], P5[QI], P6[rhyme], P7[no], P8[yes], P9[right], P10[yes], although similar results can be reported for the remaining 215 grammars. The data is drawn from a uniform distribution. As mentioned in the introduction, the error convergence to 0 corresponds to identifying the target grammar. As illustrated in Fig. 2a, the target grammar is identified within 30 - 50 iterations, irrespective of the initialisation step. Note that an exhaustive search would require "trying" all the 216 possible grammars, thus lengthening the learning process. The faster error convergence results from online nature of our learning algorithm and the incorporation of prior knowledge in our learning scenario. The algorithm is more efficient than one, where at each iteration a new grammar was selected completely at random, thus resulting in a larger number of errors and a slower convergence rate.

5.3   Varying the Probability Distributions

In the second set of experiments, we test the convergence time in two scenarios: (1) when the input data is presented to the learner from a uniform distribution, i.e. the probability to see every word is $1/n$, where n is the number of possible words produced by a given grammar; (2) certain words are more likely to occur than others. In case (2), the probability distribution is correlated with the word-length, i.e. the shorter a given word is, the higher its probability of occurrence. As can be seen in Fig. 2b, varying the probability distribution affects the convergence rate. When the data is drawn from a non-uniform distribution, the convergence rate is slower, i.e. the target grammar is identified within 80 - 150 iterations, which is still lower than the number of all the 216 possible grammars.

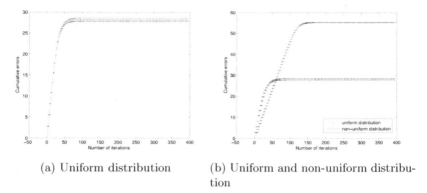

(a) Uniform distribution   (b) Uniform and non-uniform distribution

Fig. 2. Error convergence rate (with standard deviation) of the Prior Knowledge Multi-armed Bandit algorithm when the data is drawn from (a) a uniform distribution; (b) the probability distribution is correlated with the word length and compared to the uniform probability distribution

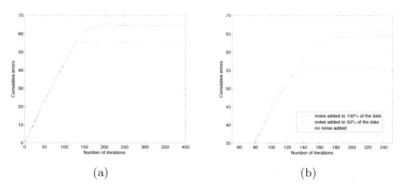

(a)   (b)

Fig. 3. (a) Error convergence rates with noise added to 0%, 50% and 100% of data. (b) expansion of Figure 3a in the critical region.

5.4   The Impact of Noise

In the third set of experiments, we added noise (ω) to the input data, or, to be more precise to the reward function. Thus, the reward was $r + \omega$, where $\omega = randn * 0.05$. $randn$ is a random value drawn from a normal distribution with mean zero and standard deviation one. We tested the influence of noise when the noise was added to varying percentage of data ranging from 0% to 100%. Fig. 3 compares the error rate convergence for cases where noise was added to 0%, 50% and 100% of data, where the data was drawn from a non-uniform distribution. The algorithm performs best with no noise present. However, even with the addition of noise, the correct grammar is identified within $110 - 170$ iterations.

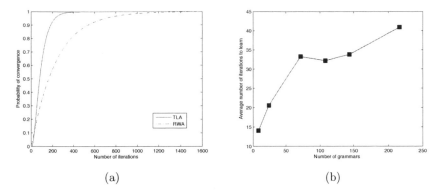

(a)   (b)

Fig. 4. (a) Probability of convergence of the Triggering Learning Algorithm (TLA) and the Random Walk Algorithm. (b) Average number of iterations required to learn the correct grammar with the prior knowledge multi-armed bandit algorithm as the size of the learning space increases.

5.5   Comparison with TLA and RWA

In the last set of experiments, we compared the performance of the 'prior knowledge multi-armed bandit" algorithm with that of TLA and RWA. Following [13], we implemented TLA as a Markov chain. Both in TLA and RWA, the words are drawn from a uniform distribution. The target grammar is the same as the one used in the previous experiments. As discussed earlier, it takes on average 30 - 50 iterations to learn the correct grammar with the prior knowledge multi-arm bandit algorithm. As can be seen in Fig. 4a, it takes 311 iterations of TLA and 1071 iterations of RWA to learn the target grammar. It must be noted that at each iteration of our algorithm the learner is given a set of 5 words, while in the case of TLA and RWA the learner is given only one word at a time. However, even if we assume the worst-case scenario, where the learner needs 50 iterations of the prior knowledge multi-arm bandit algorithm to acquire the target grammar, we still require only 250 words to learn the language. Learning with TLA and RWA requires 311 and 1071 words, respectively. The prior knowledge multi-arm bandit algorithm converges faster than TLA and RWA in spite of the fact that TLA and RWA provide the learner with additional information of transition probabilities. Note that the prior knowledge multi-arm bandit algorithm does not take into account this type of extensional information.

[12] and[13] showed that in a three-parameter setting with 8 grammars, RWA converges faster than TLA. However, our experiments on a 10-parameter space show that the convergence rate of RWA is much slower than that of TLA. We further compared the convergence rate of the three algorithms as the size of the parameter space, and consequently the number of grammars, increases. We looked at a scenario, where the number of possible grammars was: 8, 24, 72, 108, 144 and 216. As can be seen in Figs. 4b and 5b, in the case of the prior knowledge multi-arm bandit algorithm and RWA, the complexity of the learning

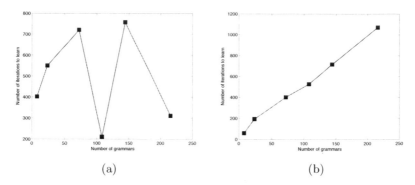

(a)   (b)

Fig. 5. (a) Number of iterations required to learn the correct grammar with TLA as the size of the learning space increases. (b) Number of iterations required to learn the correct grammar with RWA as the size of the learning space increases.

error is affected by the size of the learning space, i.e. the smaller the number of grammars the faster the learning process. The size of the learning space does not have the same effect on the TLA, i.e. there is no correlation between the number of parameters/grammars and the probability of convergence (Fig. 5a).

6   Discussion and Future Directions

The problem of learning parametrised grammars can be approached from many different perspectives. In this paper, we concentrated on the problem of error convergence, i.e. how many examples it will take the learner to reach a stage where he can parse correctly all the incoming words. We have presented a new algorithm "prior knowledge multi-armed bandit" and have shown that the algorithm can successfully tackle the problem of sample complexity. The algorithm enables the learner to acquire the target language in an online fashion without the need to resort to searching the entire parameter space and without the danger of getting stuck in a local maximum. We have also shown that the learner can "discover" the parameter setting of the target grammar without direct access to the set difference of words belonging to the different grammars, but from the more cognitively realistic access to the *Hamming distance* between the grammars parameter vectors.

A number of directions for future research arise. As the number of parameters increases, so does the complexity of the learning process. It is worth investigating how the error convergence rate will change as the parameter space grows/decreases. Further, we also need to conduct a more extensive empirical analysis of the impact of noise and data distribution on the convergence rate, i.e. how increasing the level of noise or a very unfavourable data distribution will affect the learning process.

Another possible direction is the derivation of a language change model from the current language acquisition model as well as a language acquisition model where the learner is exposed to data coming from different languages or dialects. The procedure discussed in this article concentrates on modelling the language acquisition process of a single child. Needless to say, a language change model would require scaling up the present model to an entire population.

References

1. Auer, P., Cesa-Bianchi, N., Fischer, P.: Finite-time analysis of the multiarmed bandit problem. Machine Learning (47), 235–256 (2002)
2. Bubeck, S., Munos, R., Stoltz, G., Szepesvari, S.: Online optimization in x-armed bandits. In: Proceedings of NIPS (2008)
3. Chomsky, N.: Aspects of the Theory of Syntax. MIT Press, Cambridge (1965)
4. Chomsky, N.: Lectures on government and binding. Foris, Dordrecht (1981)
5. Davis, S.M.: Syllable onsets as a factor in stress rules. Phonology (5), 1–19 (1988)
6. Dresher, B.E., Kaye, J.D.: A computational learning model for metrical phonology. Cognition 34, 137–195 (1990)
7. Gibson, T., Wexler, K.: Triggers. Linguistic Inguiry 25(4), 407–474 (1994)
8. Gold, E.M.: Language identification in the limit. Information and Control 10(4), 407–454 (1967)
9. Hale, K., Keyser, J.: On argument structure and the lexical expression of syntactic relations. In: Hale, K., Keyser, J. (eds.) The view from building 20, pp. 53–110. MIT Press, Cambridge (1993)
10. Hayes, B.: Metrical Stress Theory: Principles and Case Studies. The University of Chicago Press, Chicago (1995)
11. Kleinberg, R., Slivkins, A., Upfal, E.: Multi-Armed Bandits in Metric Spaces. In: Proceedings of STOC (2008)
12. Niyogi, P., Berwick, R.C.: A language learning model for finite parameter spaces. Cognition 61, 161–193 (1996)
13. Niyogi, P.: The Computational Nature of Language Learning and Evolution. MIT Press, Cambridge (2006)
14. Rasmussen, C.E., Williams, C.K.I.: Gaussian Processes for Machine Learning. MIT Press, Cambridge (2006)
15. Wang, Y., Audibert, J., Munos, R.: Algorithms for infinitely many-armed bandits. In: Proceedings of NIPS (2008)

Author Index